D0764136

THE CAMBRIAN FOSSILS OF CHENGJIANG, CHINA

The Flowering of Early Animal Life

Cricocosmia jinningensis, complete specimen (RCCBYU 10218), ×8.0; Mafang.

THE CAMBRIAN FOSSILS OF CHENGJIANG, CHINA

THE FLOWERING OF EARLY ANIMAL LIFE

HOU XIAN-GUANG, RICHARD J. ALDRIDGE, JAN BERGSTRÖM,
DAVID J. SIVETER, DEREK J. SIVETER AND FENG XIANG-HONG

BLACKWELL PUBLISHING
350 Main Street, Malden, MA 02148-5020, USA
108 Cowley Road, Oxford OX4 1JF, UK
550 Swanston Street, Carlton, Victoria 3053, Australia

First published 2004 by Blackwell Science Ltd
Reprinted 2004

Library of Congress Cataloging-in-Publication Data

The Cambrian fossils of Chengjiang, China: the flowering of early
animal life / Hou Xian-guang ... [et al].
p. cm.
Includes bibliographical references and index.
ISBN 1-4051-0673-5 (alk. paper)
1. Animals, Fossil—China—Chengjang (Yunnan Sheng) I. Hou,
Xian-guang.
QE756.C6C35 2004
560′.1723′095135—dc21
2003006183

A catalogue record for this title is available from the British Library.

Set in 9.5 on 12 Palatino
by SNP Best-set Typesetter Ltd, Hong Kong
Printed and bound in the United Kingdom
by The Bath Press

The publisher's policy is to use permanent paper from mills that operate a sustainable forestry policy, and which has been manufactured from pulp processed using acid-free and elementary chlorine-free practices. Furthermore, the publisher ensures that the text paper and cover board used have met acceptable environmental accreditation standards.

Fore further information on
Blackwell Publishing, visit our website:
http:/www.blackwellpublishing.com

CONTENTS

FOREWORD

The base of the Cambrian Period is one of the great watersheds in the history of life. In the earlier half of the nineteenth century, Charles Darwin had already recognized the startling change that happens in the fossil record at this horizon, when the fossil remains of metazoans appear in abundance for the first time in many localities around the world. The dawn of the Cambrian marks the appearance of mineralized shells, which apparently originated independently in several animal groups shortly after the beginning of the period. A century or more of careful collecting has only reinforced the distinctiveness of this seminal phase in the story of marine life. Initially, paleontologists concentrated on documenting the sequence of shelly fossils through the interval, in order to establish a basis for the correlation of marine strata. Trilobites—now supplemented by microfossils, like acritarchs—have proved to be of particular importance in stratigraphy for all but the lowest part of the Cambrian, and for a while our picture of early life was colored by the kind of shelly fossils that could be recovered from collecting through the average platform sedimentary rock sequence. However, there was another world that the usual fossil record did not reveal, a world of soft-bodied, or at least unmineralized, animals which lived alongside the familiar snails and trilobites, but which usually left no trace in the fossil record.

C. D. Walcott's discovery of the Middle Cambrian Burgess Shale in 1909 cast a new light upon this richer fauna. Thirty years of intensive study by several specialists at the end of the last century have made this fossil fauna one of the best known in the geological column. As well as fossils of a variety of animals that could be readily assigned to known animal phyla, the fauna included a number of "oddballs" which have stimulated much debate: were they missing links on the stem-groups of known animals, or completely new designs which left no progeny? Thanks to S. J. Gould's 1989 book *Wonderful Life*, the Burgess curiosities became well known to general readers from Manchester to Medicine Hat. But what once seemed like a unique window on to the marine world of the Cambrian has since been supplemented by other discoveries no less remarkable. Professor Hou's discovery of the Chengjiang biota in Yunnan Province, China, in 1984 proved to be a revelation equal to, or even exceeding, that provided by the fauna of the Burgess Shale. In the first place it was even older, taking us still closer to what has been described as the "big bang" at the dawn of complex animal life. Second, its preservation was, if anything, more exquisite. Third, an even greater variety of organisms was preserved—some, evidently, related to Burgess Shale forms, but others with peculiarities all of their own. The awestruck observer was granted a privileged view of a seafloor thronging with life, only (geologically speaking) a short time after the earliest shelly fossils appeared in underlying strata. The fauna included what have been claimed as the earliest vertebrates (*sensu lato*) and thus has more than a passing claim to interest in our own anthropocentric species. There are arthropods beyond imagining, "worms" of several phyla, large predators, and lumbering lobopods; while the trilobites, so long regarded as the archetypal Cambrian organism, are just one among many successful groups of animals. Once you have seen the Chengjiang fauna you will be forced to shed your preconceptions about ecological simplicity in early Phanerozoic times. This was a richly varied biota.

This book brings together marvellous color photographs to provide us with an album of Cambrian life. It is the first comprehensive "field guide" to the Chengjiang fauna. Think of it as the equivalent of one of those manuals people take to the Great Barrier Reef to identify the marine life—but here the seafloor is 525 million years old. It is a world full of surprises. The velvet worms—today represented by tropical terrestrial animals—were then much more diverse, some of them plated, spiky, odd-looking creatures. Marvel at the preservation of the comb jelly, the most delicate of marine organisms, destroyed today by the glance of an oar, but here preserved for hundreds of millions of years in almost incredible detail. Worms seem almost to be laid out upon the dissecting slab ready for inspection.

This book is much more than a mere picture gallery. The Chengjiang animals are tangible evidence of the evolutionary forcing house at the base of the Paleozoic that generated the "roots" of all the biodiversity of our living world. They bear upon fundamental questions about the generation of novelty of design. Is it feasible that so many radically different patterns of organismic construction could be derived from a common ancestor after the alleged "snowball Earth" some 600 million years ago? If so, what are the implications for genetic development? Could permutations in the expression of homeobox genes be responsible for the apparently rapid diversification of these animals? The beautiful fossil remains laid out upon their slabs of mudstone are recalcitrant facts that have to be incorporated into any scenario of molecular evolution, an invitation to further discovery, a challenge to biologists and paleontologists of the future. One thing is certain: the evidence from China will be forever built into the scientific edifice.

Or you can, if you prefer, take a delight in the aesthetic qualities of the images. You can allow your imagination to travel back to a world in which the sea swarmed with not-quite-shrimps, and the giant predator *Anomalocaris* preyed upon almost-amphioxus. Several of these animals had eyes: they looked over the living seafloor just as we contemplate the carcasses of the organisms they once observed. Paleontology is not a "dead" science. Its principal concern is to bring to life worlds that would otherwise lie forgotten and undiscovered within the rocks. It is to be hoped that this book will stimulate another generation in the quest for animals still undreamed of, vanished ecosystems still waiting to be unearthed.

Richard Fortey

PREFACE

The finding in 1984 of the Chengjiang biota, in rocks of Early Cambrian age in Yunnan Province, China, was one of the most significant paleontological discoveries of the twentieth century. The abundant and exquisite fossils, preserving fine details of the hard parts and soft tissues of animals and primitive plants approximately 525 million years old, are simply wondrous objects in their own right. More significantly, they are vital keys in helping to unravel the evolution of multicellular organisms during a period of time when such life forms first become common in the fossil record.

The Chengjiang biota is well known to practitioners and students of geology and biology through many papers published in specialist journals and in volumes resulting from scientific meetings. Much of the primary documentation is in Chinese. This is the first book in English to present an overview of the fauna, and has resulted from long established links between Professor Hou Xian-guang, the discoverer of the Chengjiang biota, and colleagues at the universities of Leicester and Oxford and the Swedish Museum of Natural History in Stockholm. The number of species known from the Chengjiang biota easily exceeds 100. Details on the authorship of each species and the date when it was established are given in the list at end of this book, together with synonyms and possible synonyms for those taxa that we are able to evaluate based on published information. It was not intended that every known species should be treated herein. We have simply provided a selection, with phyla ordered in accordance with the phylogeny of Nielsen (2001), to illustrate the range and nature of the biota. The systematic position of many Chengjiang species within their phyla is controversial and has in some cases attracted widely different opinions. Within each phylum, therefore, the order of treatment of the species does not follow any one "favored" scheme; rather, they are arranged with generally "allied" taxa together. It is hoped that with the publication of this book the sheer beauty, diversity and scientific importance of these fossils from southwestern China will become more widely known and appreciated by scientists and the public at large.

Research support underpinning this book is gratefully acknowledged from the Ministry of Science and Technology of China (G2000077702; Pandeng-95-Zhuan-01; 2002CB714007); the National Natural Science Foundation of China (No. 40272017); the Department of Science and Technology of Yunnan Province (No. 2001D0002R); the Natural Science Foundation of Yunnan Province (No. 2002D0006m); Yunnan University; the Royal Swedish Academy of Sciences; the Swedish Museum of Natural History; the Swedish Institute; the Swedish Natural Science Foundation; the Royal Society (Joint Project Q812); and the universities of Leicester and Oxford.

We thank in particular Lucy Siveter and also Chris Parks (Image Quest 3-D, Oxon, England), for invaluable assistance with computer-based photographic illustrations. Zhang Xi-guang, Ma Xiao-ya and Zhao Jie (Research Center for Chengjiang Biota, Yunnan University) kindly supplied unpublished information, and Rennison Hall (University Museum of Natural History, Oxford) and Michael Lear (Filey, North Yorkshire, England) provided technical help. We are grateful to Pollyanna von Knorring and Javier Herbozo (Swedish Museum of Natural History) for their skill in drawing many of the reconstructions of the fossils featured in this book. We are indebted to Professor Derek Briggs (Yale University) for his constructive review of the manuscript. Thanks also to the staff at, or those associated with, Blackwell Publishing (Oxford) who were involved in bringing this book to fruition: Ian Francis for accepting the outline proposal, Delia Sandford the managing editor, and Jane Andrew, Cee Brandson, Jo Egré, and Rosie Hayden for their various editorial inputs.

All of the Chengjiang specimens figured in this book are from the Lower Cambrian Yu'anshan Member, Heilinpu (formerly Qiongzhusi) Formation, of Yunnan Province. Certain specimens have been illustrated previously in the scientific literature, but one of us (Derek J. S.) re-photographed some of these for this book, together with very many other specimens that are figured herein for the first time. The figured Chengjiang material is housed at the Research Center for the Chengjiang Biota, Yunnan University (RCCBYU), and the Nanjing Institute of Geology and Paleontology (NIGPAS), Academia Sinica, the People's Republic of China.

PART ONE

GEOLOGICAL AND EVOLUTIONARY SETTING OF THE BIOTA

1 GEOLOGICAL TIME AND THE EVOLUTION OF EARLY LIFE ON EARTH

Our planet is some 4,600 million years old. We have no direct record of Earth history for the first 700 million years or so, but rocks have been found that date at least as far back as 3,800 million years, and perhaps even more than 4,000 million years. Earth history has been divided up into three eons: the Archean, the Proterozoic, and the Phanerozoic (Fig. 1.1); the Archean and Proterozoic are jointly termed the Precambrian. The boundary between the extremely ancient Archean and the very ancient Proterozoic is drawn at 2,500 million years, while the beginning of the Phanerozoic (literally meaning "manifest life") is recognized by evolutionary changes shown by fossil animals about 545 million years ago. The Proterozoic is divided into three periods, the Paleoproterozoic (2,500–1,600 million years), the Mesoproterozoic (1,600–1,000 million years), and the Neoproterozoic (1,000–545 million years). The earliest period of the Phanerozoic eon is called the Cambrian, after the old Latin name for Wales, and it was during this time that almost all the animal groups we now know on Earth made their initial appearances. Some of the most important fossil evidence for these originations has come from the Chengjiang biota of southwest China — the subject of this book.

The record of life on Earth, however, goes much further back in time, perhaps nearly as far as the record of the rocks. Possible microfossils that resemble cyanobacteria have been reported from rocks as old as 3,500 million years in Australia (Schopf 1993) and there is circumstantial evidence from geochemical studies that carbon isotopes were being fractionated by organic processes as long ago as 3,860 million years (Mojzsis *et al.* 1996). However, there is a need to treat these reports of evidence for very early life with caution, and the further back in time the record is extended the more controversial the claims become (see, for example, Brasier *et al.* 2002, Fedo & Whitehouse 2002, van Zuilen *et al.* 2002).

It is quite probable, however, that fossils are present in rocks of Archean age, albeit extremely rarely. All of these sparse organic remains are microscopic, sometimes filamentous, and may be associated with laminated sedimentary structures known as stromatolites (Fig. 1.2a). Modern stromatolitic structures are built up through microbial growth, with successive layers of sediment being trapped by the microbial mats. The resulting forms are commonly dome-like or columnar, and these characteristic shapes can also be recognized in Archean sediments up to 3,500 million years in age. Once again, the very oldest stromatolites are somewhat controversial, and it is possible that they could have been constructed by abiotic processes rather than by living organisms (Grotzinger & Rothman 1996).

The micro-organisms identified living in modern stromatolitic communities represent a wide range of types of life, including filamentous and coccoid cyanobacteria, microalgae, bacteria, and diatoms (Bauld *et al.* 1992). It is normally considered that most Precambrian examples were simpler, primarily constructed by the "blue-green" cyanobacteria, although the very earliest may have been built by anaerobic photosynthetic bacteria (Walter 1994). If we accept the combined evidence from putative microfossils, stromatolites and carbon isotopes, then it appears that life may have begun on Earth as much as 3,500 million years ago and that these life forms may have included micro-organisms that could generate their own energy by photosynthesis. But it is certain that the debate surrounding the nature of the earliest traces of life is far from over.

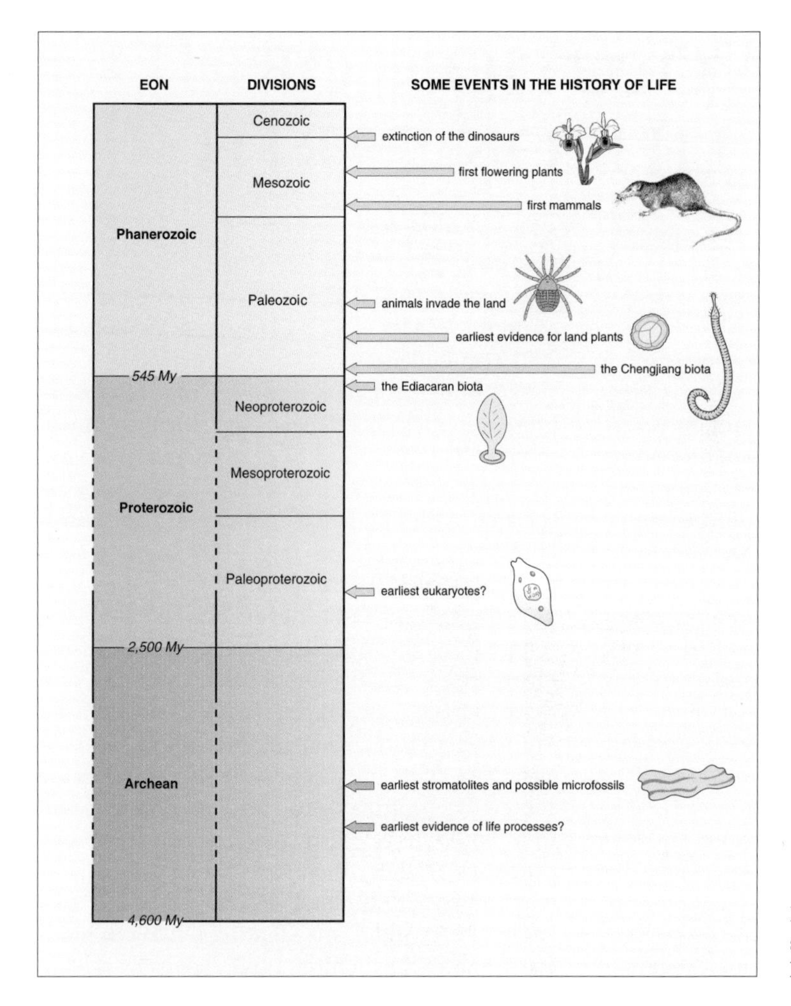

Figure 1.1 The geological time scale, showing some major events in the history of life.

There is a much richer and less controversial preserved record of life in strata of Paleoproterozoic and Mesoproterozoic age. Microbial mats and stromatolites constructed by cyanobacteria are quite abundant, and it is likely that cyanobacteria had become diversified by the mid-Paleoproterozoic (Knoll 1996). Geochemical evidence also indicates that oxygen levels in the atmosphere had begun to build up, reaching about 1% of the current level. There are also biomolecular data showing that one of the most significant steps in evolutionary history had taken place by this time—the appearance of eukaryotic cells (Brocks *et al.* 1999). Eukaryotes are distinguished from the more primitive prokaryotes by their larger size, and by their much more complicated organization with a discrete nucleus, containing DNA organized on chromosomes, and a variety of organelles within the cytoplasm. Simple algal cysts with a cell size larger than any known in modern prokaryotic organisms date back to about 2,100 million years (Han & Runnegar 1992). Somewhat more complex spherical and spinose forms ("acritarchs", Fig. 1.2b), presaging a diversification of eukaryotic micro-organisms, have been reported from rocks in northern Australia nearly 1,500 million years old (Javaux *et al.* 2001). Records also indicate that, by the late Mesoproterozoic, a

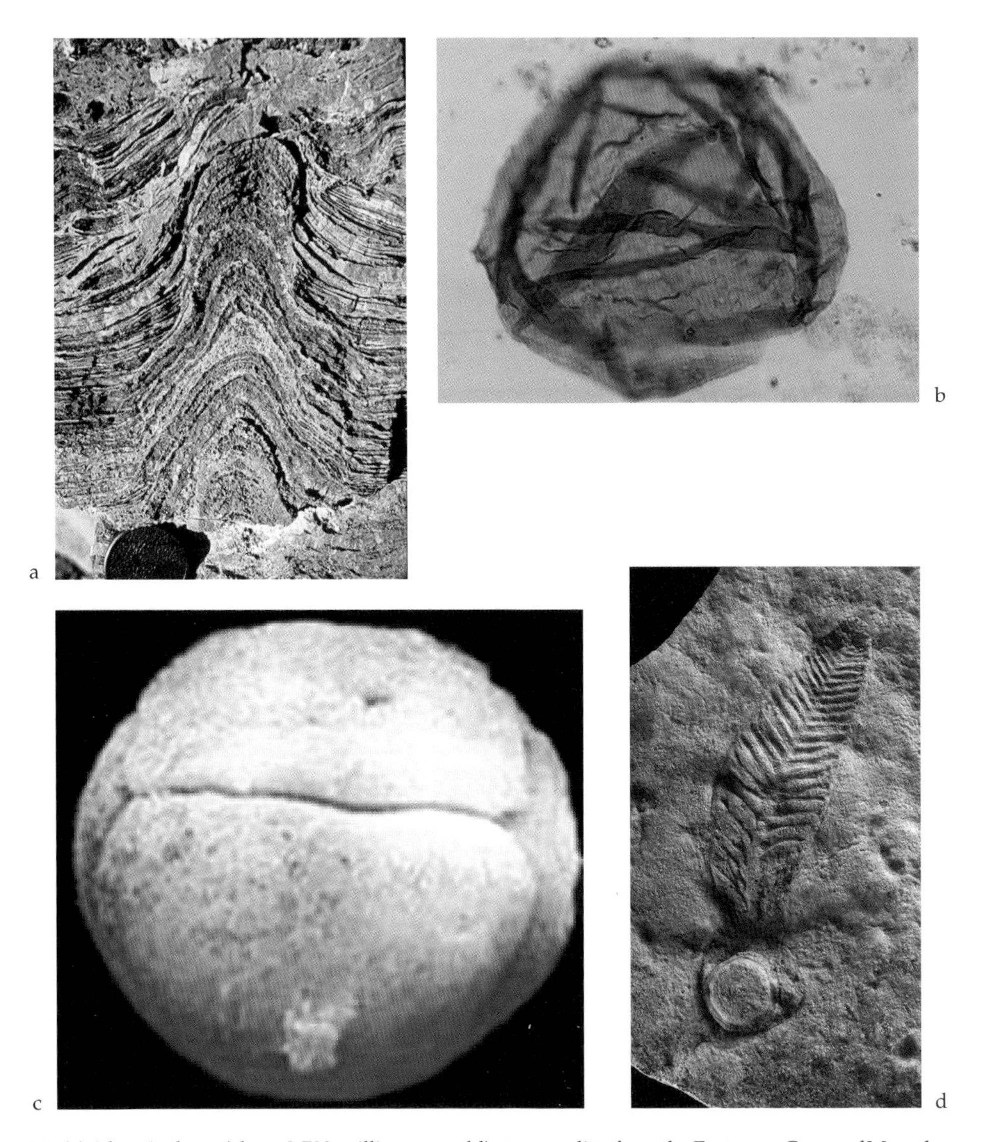

Figure 1.2 (a) A late Archean (about 2,700-million-year old) stromatolite, from the Fortescue Group of Meentheena, Western Australia; lens cap for scale (courtesy of Clive A. Boulter). (b) A Proterozoic acritarch (about 1,400 million years old), from the Roper Group, Northern Territory, Australia; width of specimen about 50 μm (courtesy of Christopher A. Peat). (c) A fossil embryo from the Precambrian Doushantuo phosphorite of south China; width of specimen 505 μm (courtesy of Xiao Shu-hai). (d) *Charniodiscus arboreus* (Glaessner, 1959), from the Ediacaran biota of South Australia (South Australian Museum specimen SAM P19690a); length of specimen 40 cm (courtesy of A. Bronikov and Dimitri Grazhdankin).

variety of groups of red and green algae had appeared.

This diversification of eukaryotes continued through the Neoproterozoic. Exceptionally well-preserved specimens in the Doushantuo Formation of South China show that by about 590 million years ago multicellular algae were highly diverse and that the communities probably included brown algae (Xiao *et al.* 1998, 2002). Particularly remarkably, phosphorites from the same formation have yielded embryos, probably of animals, three-dimensionally preserved in the early stages of cleavage (Xiao *et al.* 1998, Xiao 2002). The embryos are consistently about half a millimeter in diameter, compartmentalized into two, four, eight, or more internal bodies (Fig. 1.2c). Associated with them are tiny sponges, with monaxonal (single-axis) spicules and preserved soft tissues (Li *et al.* 1998). This incredible assemblage of fossils may record some of the earliest stages of the evolution of metazoan animals.

Metazoa are characterized by the grouping together of numerous cells, with different sets of cells fulfilling different specialized functions. The first metazoans must have arisen during the Proterozoic from a single-celled eukaryotic ancestor, but there is considerable debate about the timing of this event. Molecular clock calculations, which assume a regular substitution rate within selected genes, suggest that divergence of the metazoans occurred more than 1,000 million years ago (Wray *et al.* 1996), but there is no widely accepted fossil record of multicellular animals older than about 600 million years. In fact, the earliest undoubted metazoan body fossils are of about the same age as the Doushantuo embryos. These are part of the Ediacaran biota, named after sites in the Ediacara Hills of South Australia, where diverse macrofossils have been extensively studied.

Ediacaran fossils are now known from more than 30 localities worldwide. Among the earliest fossils that have been allocated to this assemblage are unornamented discs and rings that have been found in the Mackenzie Mountains of Canada (Narbonne *et al.* 1994), although these cannot be grouped unequivocally with the more diverse forms of the true Ediacaran biota (Fig. 1.2d). The specimens from the Mackenzie Mountains are dated as slightly older than 600 million years, and a somewhat greater variety of forms, including discs, triangles and fronds, has been reported from strata in Newfoundland that are 570–580 million years in age (Narbonne & Gehling 2002). The main Ediacaran organisms, however, are found in rocks spanning an interval from about 565 to 543 million years ago, immediately above tillites that record the most extensive glacial episode in Earth history. Many workers have related the variety of soft-bodied forms found in these strata to well-known animal phyla, principally the Cnidaria, Annelida, Mollusca, Arthropoda, and Echinodermata. Although sponges, cnidarians and molluscs are almost undoubtedly represented, Seilacher (1992) proposed that most of these fossils belonged to a distinct and independent clade, the Vendobionta, with a construction like an air mattress and totally different from that of subsequent animals. One commentator has even suggested that they are not animals at all, but represent a range of lichens (Retallack 1994). More recently, Seilacher *et al.* (2003) have argued that the vendobionts were gigantic unicellular organisms living within biomats as part of an ecological system consisting almost entirely of primary producers and decomposers. Whatever their relationships, the Ediacaran taxa almost all disappeared by the beginning of the Cambrian, with just a few specimens in Cambrian strata suggesting that these forms persisted for a while alongside their more familiar successors. If Seilacher and his co-workers are correct, the extinction of the Vendobionta was followed by an immediate radiation of metazoan animals that, until that time, had been a minor component of the biota. The concomitant development of hard parts instigated more complex trophic webs, with carnivores, scavengers, and consumers diversifying in the Early Cambrian seas.

One open question at the moment is whether the Doushantuo embryos are the early life stages of Ediacaran organisms, or whether they represent other taxa. Most commentators have compared them to the embryos of nematode worms, flatworms, arthropods, or cnidarians, but this in itself does not provide an answer.

Other evidence of animal life in the Neoproterozoic comes from trace fossils. Several of these, particularly the earlier ones, are controversial, but trace fossil assemblages are clearly recognizable in strata coeval with the Ediacaran biotas, and in strata that are perhaps slightly older than the diverse Ediacaran assemblages. Mostly, these traces are simple tracks and horizontal burrows, with some meandering grazing structures, but there appears to have been insufficient activity to cause complete reworking (bioturbation) of the sediment. The animals responsible for these traces are not normally preserved as fossils (at least not so that the link between the two can be demonstrated), but the trails are generally attributed to the activities of mobile bilateral "worms" with hydrostatic skeletons.

There are recent claims of much older trace-like fossils from Western Australia, where parallel pairs of ridges, straight or curved, occur in rocks dated as more than 1,200 million years old (Rasmussen *et al.* 2002). Intriguingly, these traces are associated with discoidal imprints that are possibly of biogenic origin and might represent the earliest Ediacara-type fossils discovered to date. This discovery serves to illustrate the fact that new finds of Proterozoic fossils are being reported at an ever-increasing rate, and it is clear that scenarios for the evolution of Precambrian life are facing continuing modification as these new data appear.

2 THE EVOLUTIONARY SIGNIFICANCE OF THE CHENGJIANG BIOTA

The Cambrian bears witness to a remarkable increase in the abundance, types and groups of fossils in the rock record, a feature considered by many to represent a major radiation in the evolution of life, the so-called "Cambrian Explosion". In essence all of the major groups of life that are known from the present day can be traced back to the Cambrian. This is captured not only in the earliest fossil evidence of the hard skeletal parts of animals but also by the occurrence of several exceptionally preserved fossil biotas (Konservat-Lagerstätten), in which the soft parts of animals and entirely soft-bodied life forms are represented. The Upper Cambrian microarthropod-dominated "Orsten" fauna of Sweden (Müller 1979; see references in Walossek & Müller 1998), the celebrated Middle Cambrian Burgess Shale assemblage of British Columbia (e.g. Briggs *et al.* 1994, Conway Morris 1998) and the Lower Cambrian biotas from Sirius Passet in Greenland (e.g. Conway Morris & Peel 1995, Budd 1998) and from Chengjiang, are the key Cambrian Konservat-Lagerstätten. The Chengjiang biota (Fig. 2.1) is exceptional in providing a comprehensive and very early window on the nature of Phanerozoic life.

Although the first appearance of many forms of multicellular life in the Cambrian has traditionally been taken to indicate a major diversification of life (see Gould 1989), the evolutionary events that gave rise to the metazoans possibly occurred deep in Proterozoic time (Fortey *et al.* 1996, 1997, Conway Morris 1997b, Walossek 1999; see also Bergström 1991, 1994). Thus, the nature and validity of the "Cambrian Explosion" attracts intense debate (see Wills & Fortey 2000). Some authors conclude that the explosion is real (Conway Morris 2000a). Some consider that it is a false explosion, perhaps a preservational window when, *inter alia*, animals became larger and acquired the hard parts that offered greater fossilization potential (Fortey *et al.* 1996, 1997). In support of the latter opinion it has been argued that the presence in the Early Cambrian of relatively advanced forms of, for example, arthropods (e.g. Siveter *et al.* 2001a), implies that less derived members of the phylum and other less derived metazoan groups must have differentiated still earlier in time (Fortey *et al.* 1996, 1997, Fortey 2001; see Siveter *et al.* 2001b). Other researchers have stressed that the appearance of body plans familiar from modern animals ("crown groups" of their terminology) mostly appeared after the Early Cambrian (Budd & Jensen 2000, Budd *et al.* 2001; see also Conway Morris 2000b) and favor a model of progressive metazoan diversification through the end of the Proterozoic to beyond the Cambrian. Whatever the truth of what happened to life across the Precambrian–Cambrian boundary and why, it is undisputed that the body fossils and trace fossils of the Chengjiang biota offer unparalleled insight into biodiversity, paleobiology, autecology and community structure for a crucial period in the history of life.

Figure 2.1 A cache of Chengjiang fossils from near Haikou: the arthropods *Fuxianhuia* (lower right) and *Eoredlichia* (a trilobite, lower left) and various worms (top left), ×1.1.

3 THE DISCOVERY AND INITIAL STUDY OF THE CHENGJIANG LAGERSTÄTTE

The Chengjiang Lagerstätte was discovered by Hou Xian-guang in 1984 and is now known from many localities of the Yu'anshan Member, Heilinpu Formation, over a wide area of eastern Yunnan Province. The first soft-bodied fossils to be found were from Maotianshan (Maotian Hill; Fig. 3.1), about 6 km east of the county town of Chengjiang. The Kunming-Chengjiang area of Yunnan Province is one of the best-known geological areas of China. From the pioneer days of geological exploration in China it had been appreciated that the Lower Cambrian of eastern Yunnan Province is richly fossiliferous. As early as the first decade of the twentieth century the Frenchmen Honoré Lantenois (1907), Jaques Deprat (1912), and Henri Mansuy (1907, 1912) studied the geology and paleontology of the region (Figs 3.2–3.4), resulting in publications that featured new fossils including trilobites and other arthropods. As part of mapping and other general geological survey work, the Lower Cambrian in this area was also extensively studied in the 1930s and 1940s (see Babcock & Zhang 2001, Hou *et al.* 2002b). Indeed, the sequence has long been taken as a standard for the stratigraphic subdivision and correlation of the Cambrian, not only within the Southwest China (Yangtze) Platform but also throughout China and beyond.

In June 1984 Hou Xian-guang, then a member of the Nanjing Institute of Geology and Paleontology of the Chinese Academy of Sciences, arrived in Kunming City to begin his second stint of fieldwork for his research on bradoriid arthropods (Hou *et al.* 2002b). Already in 1980 he had systematically collected bradoriids at the Qiongzhusi section in Kunming City and from Sichuan Province. That the Kunming-Chengjiang area is especially rich in bradoriids was elucidated much earlier, by Professor Yang Zui-yi, during the 1930s (see Ho 1942). As a consequence of hostilities within China, Yang's Department of Geology at Zhongshan University had moved from Guangzhou City in Guangdong Province to the village of Donglongtan, situated about 55 km southeast of Kunming and a mere 1.5 km west of Maotianshan. Following fieldwork in Jinning County southwest of Kunming, Hou Xian-guang had travelled to Chengjiang town and then on by cart to the nearby small village of Dapotou, where a team from the Geological Bureau of Yunnan Province was living, prospecting for phosphorite deposits in the Lower Cambrian. After reviewing the Heilinpu Formation at several nearby localities, systematic collection of bradoriids from near Hongjiachong village was undertaken with the help of a hired farm worker, but the sequence was demonstrably incomplete and a section on the west slope of Maotianshan was finally selected for detailed study (Fig. 3.5).

The mudstone blocks that the farm worker dug out at Maotianshan were scoured for bradoriids. Work was notably easier than at Dapotou and Hongjiachong, because the rock was strongly weathered. At about three o'clock in the afternoon of Sunday July 1, a semicircular white film was discovered in a split slab, and was mistakenly thought to represent the valve of an unknown crustacean. With the realization that this and a second, subelliptical exoskeleton represented a previously unreported species, breaking of the rock in a search for additional fossils continued apace. With the find of another specimen, a 4–5 cm

Figure 3.1 Maotianshan, near Chengjiang. A section on the facing (west) slope of the hill, below the wooded area, where the Chengjiang biota was discovered.

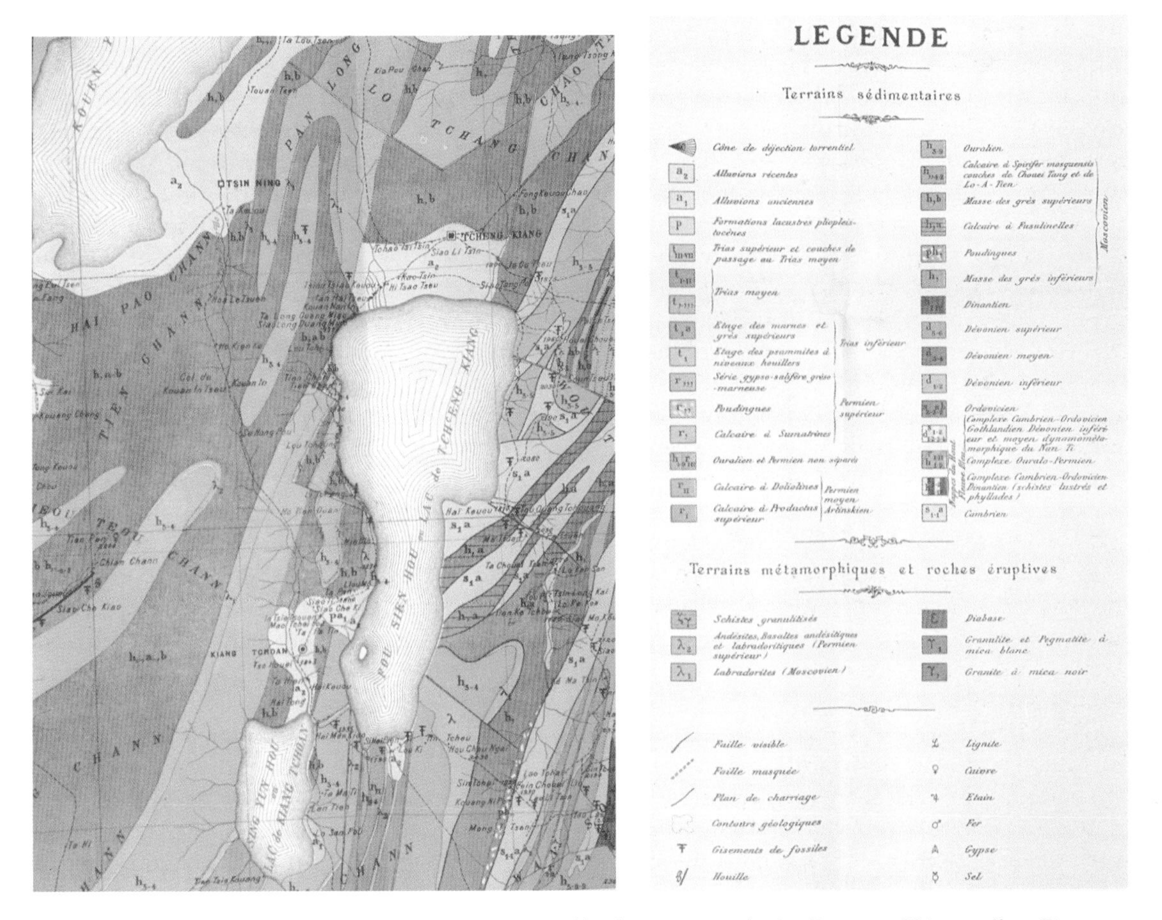

Figure 3.2 The geology of the Chengjiang area, as recognized by the pioneer geologists Deprat and Mansuy (from Deprat 1912). Fuxian Lake, the much smaller Xingyun Lake and the southern end of Dianchi Lake are also shown.

Figure 3.3 View across the plain in the vicinity of the county town of Chengjiang (from Deprat 1912).

Figure 3.4 View looking north along the north-western shore of Fuxian Lake (from Deprat 1912).

Figure 3.5 Hou Xian-guang at the locality at Maotianshan, where he discovered the first soft-bodied fossils of the Chengjiang biota.

Figure 3.6 The specimen that signaled the discovery of the Chengjiang biota: the soft-bodied arthropod *Naraoia longicaudata*, ×2.0.

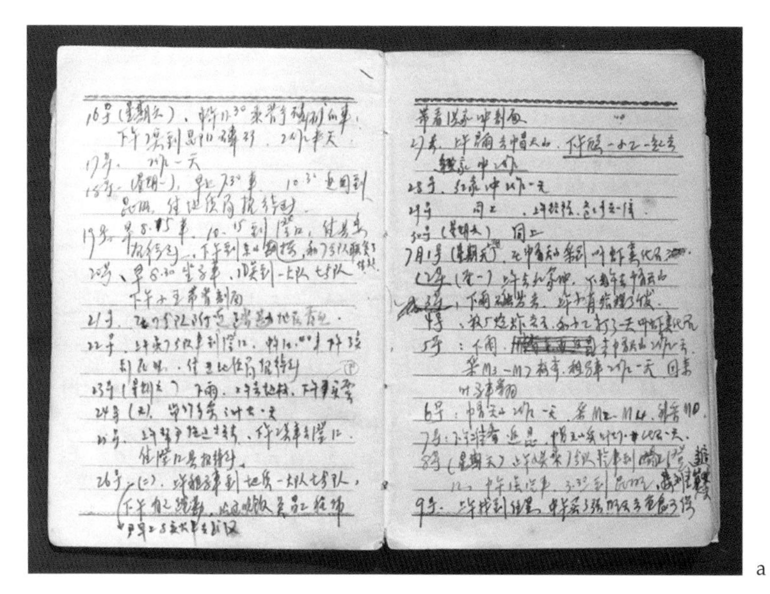

a

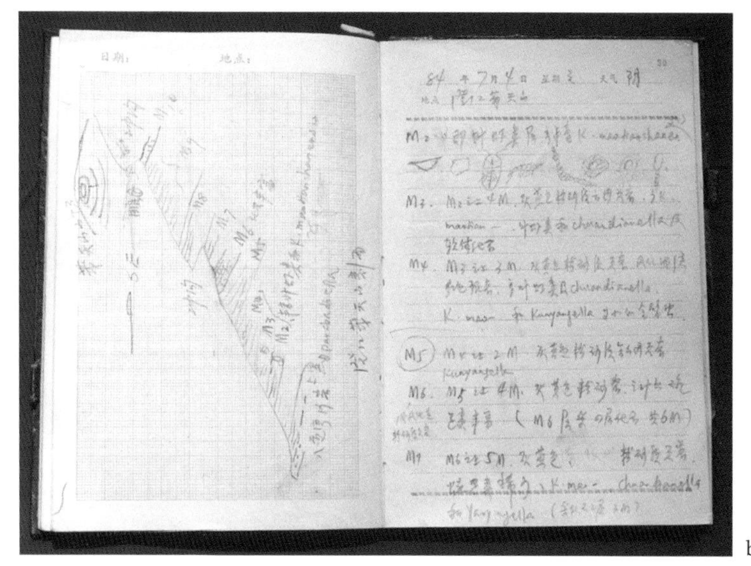

b

Figure 3.7 (a) Hou Xian-guang's field diary, recording the day that the soft-bodied fossils of the Chengjiang biota were discovered at Maotianshan, and (b) his field notebook, with drawings of fossils and sketches and notes on the geology of the section.

long animal with limbs preserved, it became apparent that here was nothing less than a soft-bodied biota. As recalled by Hou Xian-guang, the specimen with appendages (subsequently selected as the holotype of the arthropod *Naraoia longicaudata*) (Fig. 3.6) appeared as if it was alive on the wet surface of the mudstone. Elated by the discovery, the searchers increased their efforts and other new soft-bodied fossils were revealed one after another. Work on the section did not end

Figure 3.8 The search for soft-bodied fossils: splitting rock at Maotianshan.

Figure 3.9 View from Xiaolantian village towards the eponymous locality at the top of the track on the opposite side of the valley.

until dark. Hou's field diary for that day signaled the significance of his discovery by alluding to the Burgess Shale fauna: "The discovery of fossils in the Phyllopod Bed" (Fig. 3.7).

Members of the Geological Bureau team continued to provide Hou Xian-guang with valuable assistance in the field, not least by blasting the trackside exposures of the west slope of Maotianshan. Soft-bodied and other fossils were then collected on a large scale from three broad stratigraphic levels that were designated M2 (oldest), M3 and M4 (youngest) respectively. These terms were subsequently applied to three small quarries that were opened up for collecting. The three levels correspond to at least ten beds (Hou 1987a), but in fact it is almost impossible to determine exactly how many beds of the blocky mudstone bear soft-bodied fossils. The mudstone of level M2 is 5 m thick and yields many species of the Chengjiang fauna; in general the number of taxa and specimens successively decreases through levels M3 and M4.

Hou Xian-guang also undertook a systematic search for bradoriids at several other sections in Yunnan Province, for example at Meishucun in Jinning County, Sapushan (Sapu Hill) and Shishan (Shi Hill) in Wuding County, and Kebaocun in Yiliang County, as well as at Hongjiachong and Dapotou in Chengjiang County. In addition to bradoriids, these sections yielded specimens of soft-bodied and lightly sclerotized and mineralized animals such as worms (*Cricocosmia*), large bivalved arthropods (e.g. *Isoxys*), a brachiopod (*Heliomedusa*), and an isolated sclerite of a lobopodian (*Microdictyon*), more specimens of which were subsequently obtained from Meishucun in 1986 (Hou & Sun 1988). These 10 weeks of fieldwork, ending on August 17, 1984, had demonstrated that fossils with soft-part preservation are widely distributed in eastern Yunnan Province, and that in order to obtain reasonable numbers of specimens it is necessary to split large amounts of rock (Fig. 3.8). Letters sent by Hou from the field, in 1984, informed the directors and others at his institute in Nanjing about the collection of abundant, well-preserved bradoriid specimens (in part treated in Hou 1987d), his finds of the oldest trilobites at Chengjiang, Wuding and

Figure 3.10 Collecting at Xiaolantian, with the village in the background.

Figure 3.11 View from the section at Ma'anshan.

Jinning (some material reported by Zhang 1987a) and other trilobites from Maotianshan (Zhang 1987b), and the discovery and collection of many fossils with preserved soft parts (e.g. Zhang & Hou 1985, Hou 1987a, 1987b, 1987c, Sun & Hou 1987a, 1987b).

Hou Xian-guang's subsequent fieldwork in the Chengjiang area was specifically aimed at collecting fossils with soft-part preservation. By the time of his next visit, from April to June 1985, logistics had changed. The team from the Geological Bureau of Yunnan Province resident in Dapotou village had new leaders; the ground on both sides of the cart road leading from Dapotou to Maotianshan had been cleared to provide better access to two new phosphorite factories; and a drilling group of the Geological Bureau of Yunnan Province was living at the foot of Maotianshan and offered generous assistance, especially in providing provisions in the field.

With the support of the directors of the Nanjing Institute of Geology and Paleontology, a third field season was undertaken throughout October to December 1985. For part of this period Hou Xian-guang was joined by Chen Luan-sheng, the custodian of fossils at the museum in the Nanjing Institute. From April to September 1987, further large-scale collecting took place, again supported by Academia Sinica, when work was concentrated mainly at Maotianshan and Jianbaobaoshan near Dapotou village. Hou's colleagues, Chen Jun-yuan, Zhou Gui-qing and Zhang Jun-ming, joined the field group at that time but they left in early May and June respectively for other duties. Additional collections were made by Hou Xian-guang in November 1989 and April–May 1990, especially from new sections such as those at nearby Fengkoushao, Xiaolantian (Figs 3.9, 3.10) and Ma'anshan (Fig. 3.11).

The initial phase of collecting and describing the Chengjiang biota ended when Hou Xian-guang left China for a lengthy period of research cooperation with Swedish scientists at the Natural History Museum in Stockholm, endeavours that resulted in papers on a wide range of Chengjiang animals (see Hou & Bergström 1997 and references therein). Subsequent to the early phase of study, Chengjiang fossils also engaged the attention of many Chinese and other paleontologists worldwide, generating numerous research publications. In particular, Chen Jun-yuan (Nanjing Institute of Geology and Palaeontology) and Shu De-gan (North-west University, Xian) and other Chinese scientists and their collaborators have made considerable additions to the literature on the biota (see References).

4 THE DISTRIBUTION AND GEOLOGICAL SETTING OF THE CHENGJIANG LAGERSTÄTTE

The Cambrian of eastern Yunnan Province is part of the Southwest China Platform. During Early and Mid-Cambrian times, southwest China formed part of a discrete South China Plate lying at low latitudes and marginal to the Gondwana paleocontinent (McKerrow *et al.* 1992; Fig. 4.1). The area lay within the Redlichiid trilobite biogeographical realm, the faunas of which also characterize the Cambrian of north China, Australia, Southeast Asia and parts of central Asia (Pillola 1990). To the west, the Cambrian faunas of North America (including Greenland), southern Britain and parts of Scandinavia, respectively representative of the faunas of the ancient continents of Laurentia, Avalonia and Baltica, were characterized by trilobites that distinguish the Olenellid Realm. Areas between the Redlichiid and Olenellid realms, such as Siberia and North Africa, are distinguished by trilobites of the Bigotinid Realm.

The western part of the Southwest China Platform, the so-called Western Subprovince (Fig. 4.2), includes Lower Cambrian rock in southern Shaanxi and eastern Sichuan provinces and in various counties and Kunming City of eastern Yunnan Province. With one possible exception, the Chengjiang Lagerstätte occurs exclusively in the Yu'anshan Member, the younger of two members of the Lower Cambrian Heilinpu Formation (Fig. 4.3). Sediments of the Heilinpu Formation accumulated in shallow marine conditions and overlie thick phosphorites of the upper part of the Yuhucan Formation. The phosphogenesis is associated with a late Neoproterozoic to Early Cambrian transgression that resulted in the flooding of the Southwest China Platform (see Siegmund 1997, Babcock & Zhang 2001).

The Lower Cambrian of eastern Yunnan is zoned principally on trilobites and, in the lower part of the sequence, small shelly fossils (Fig. 4.3). The Chengjiang Lagerstätte occurs in a biozone characterized by the trilobite genera *Eoredlichia* and *Wutingaspis*, which succeeds the *Parabadiella* Biozone, so named after the oldest known trilobite in Yunnan Province. Based on a chain of correlation involving mostly small shelly fossils, trilobites, and also acritarchs, the Chengjiang Lagerstätte is generally considered to correlate with the Atdabanian Stage in Siberian terms, although some evidence suggests a slightly younger, Early Botomian age (see Hou & Bergström 1997, Zhang *et al.* 2001). The biota is therefore slightly older than, or perhaps coeval with, the Sirius Passett Lagerstätte from the Lower Cambrian of Greenland and is approximately 525 million years old.

The Yu'anshan Member occurs widely in a sedimentary basin that extends over tens of thousands of square kilometers and the Chengjiang biota is now known from many areas of eastern Yunnan Province (Fig. 4.4). In the years immediately following Hou's 1984 discovery, collecting focused on Maotianshan and several other nearby localities in Chenjiang County, such as Hongjiachong, Dapotou, Fengkoushao and especially Xiaolantian and Ma'anshan. At the same time, soft-bodied fossils were also found further afield, at Meishucun in Jinning County, Kebaocun in Yiliang County, and Sapushan and Shishan in Wuding County (Luo & Zhang 1986, Hou & Sun 1988, Hou & Bergström 1997). Perhaps some of the most important new localities to furnish the Chengjiang fauna are at Ercaicun and Mafang

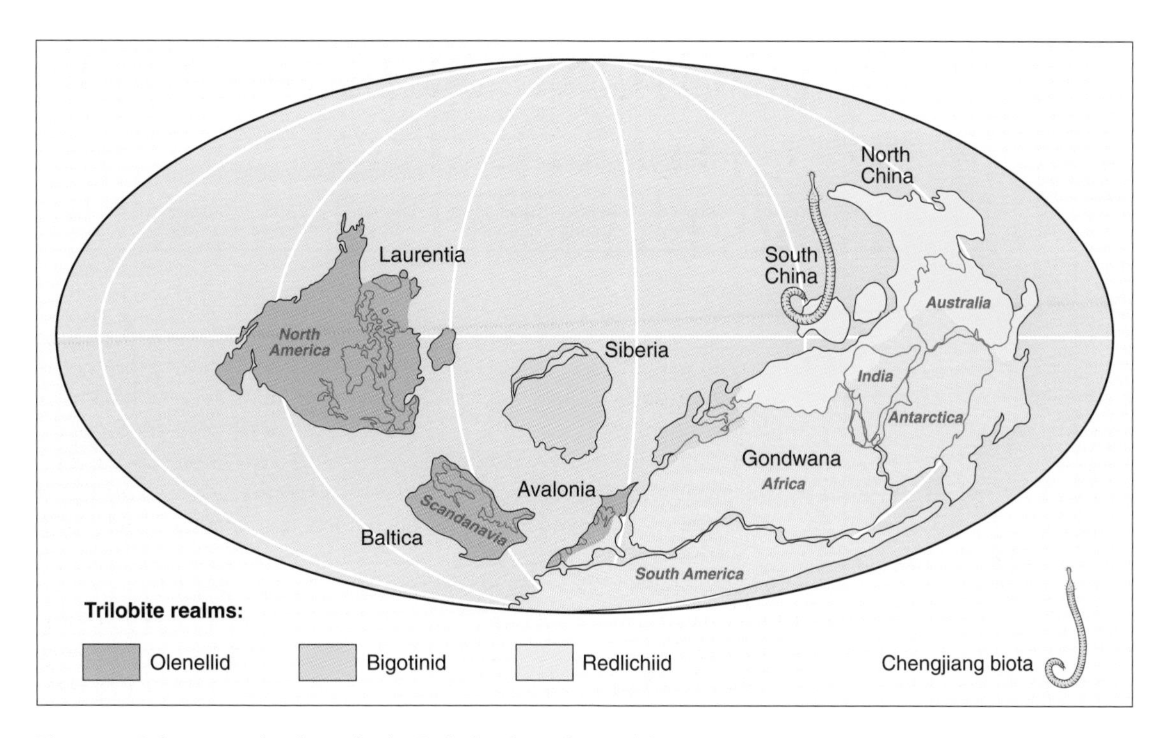

Figure 4.1 Paleogeographical map for the Early Cambrian (late Atdabanian–Toyonian stages), showing trilobite faunal realms (after McKerrow *et al*. 1992; trilobite faunal realms from Pillola 1990).

(Figs 4.5, 4.6), near Haikou town, close to the west side of Dianchi Lake (Luo *et al*. 1997). Finds from Ercaicun include the world's oldest known vertebrates (Shu *et al*. 1999, 2003, Hou *et al*. 2002a). Sections in Anning, Wuding, Malong and Yiliang areas, and at Dahaiyan near the east shore of Dianchi lake, have yielded additional soft-bodied and lightly sclerotized Chengjiang fossils (Zhang *et al*. 2001). Specimens from Kanfuqing village, in Malong County, include appendages of the large carnivore *Anomalocaris* and are purported to be from the Canglangpu Formation, which overlies the Heilinpu Formation.

The Shiyantou Member of the Heilinpu Formation is essentially a gray to black siltstone-dominated unit. The succeeding, Lagerstätte-bearing Yu'anshan Member ranges from 100 to 150 m in thickness and consists mostly of gray to greenish-gray finely-bedded mudstones that weather to a characteristic yellowish hue, together with silty and sandy interbeds that increase in frequency in the higher horizons (Fig. 4.3). Sediments of the Yu'anshan Member are generally considered to have accumulated in shallow marine conditions, but there are several ideas regarding the detailed nature of the depositional setting and processes involved. Some consider that the upward coarsening sequence of the Heilinpu and Canglangpu formations formed in association with a prograding delta, constructed at one or more river mouths (Chen & Zhou 1997). In this scenario the mudstones yielding the soft-bodied biota are interpreted as distal marine deposits that in part reflect episodic turbidity current activity (Lindström 1995), and the upper, more sandy and silty levels are said to be prodelta deposits that accumulated prior to the delta-front (basal Canglangpu) sands. Others maintain that

SYSTEM	CHINESE STAGE	CHINESE BIOZONE	WESTERN SUBPROVINCE OF SOUTHWEST CHINA (YANGTZE) PLATFORM							
			YUNNAN PROVINCE				CENTRAL SICHUAN PROVINCE		NORTHERN SICHUAN	SOUTHERN SHAANXI
			Kunming City	Wuding County	Jinning County	Chengjiang County	Leshan City	E'mei County	Guangyuan City	Hanzhong City
CAMBRIAN (LOWER)	Longwang-miaoian	*Redlichia guizhouensis*	Longwangmiao Formation				Longwang-miao Formation	Taiyangping Formation		
		Hoffetella								
	Canglangpuian	*Megapalaeolenus*	Canglangpu Formation	Wulongjing Member			Canglangpu Formation	Yuxiansi Formation		?
		Palaeolenus								Unexposed
		Sichuanolenus -Paokannia		Guanshan Member						
		Metaredlichioides -Chengkouia								?
		Drepanuroides				Canglangpu Formation?			Modaoya Formation	
		Yunnanaspis -Yiliangella								
	Qiongzhusian	*Yunnanocephalus -Malungia*	Heilinpu Formation	Yu'anshan Member				Jiulaodong Formation	Qiongzhusi Formation	Qiongzhusi Formation
		Eoredlichia -Wutingaspis								
		Parabadiella								
	Meishucunian	*Sinosachites -Tannuolina*		Shiyantou Member						
		Barren interval	Yuhucun Formation				Maidiping Formation		?	?
		Heraultipegma								
		Paragloborilus -Siphogonuchites								
?		*Anabarites -Protohertzina*								
PRE-CAMBRIAN	Dengying -xian	No zones defined								

Figure 4.2 Generalized stratigraphy of the upper Precambrian to Lower Cambrian of the Western Subprovince of the Southwest China Platform (modified from Hou *et al.* 2002b, with additions from Qian *et al.* 2001). The arrow indicates the stratigraphic position of the Chengjiang biota.

the biota-rich mudstones show alternating millimeter-scale lighter and darker couplets, a periodicity that possibly results from tidally influenced sedimentation, and that the coarser interbeds probably reflect frequent episodes of higher energy, storm-influenced conditions (Babcock *et al.* 2001, Babcock & Zhang 2001). Poly-cyclic anoxia and rapid burial of the biota are processes suggested in other depositional models (Chen & Erdtmann 1991).

In a recent study, aimed at more securely furnishing a depositional model for the sediments containing the Chengjiang biota, detailed logging of the Yu'anshan Member at Maotianshan

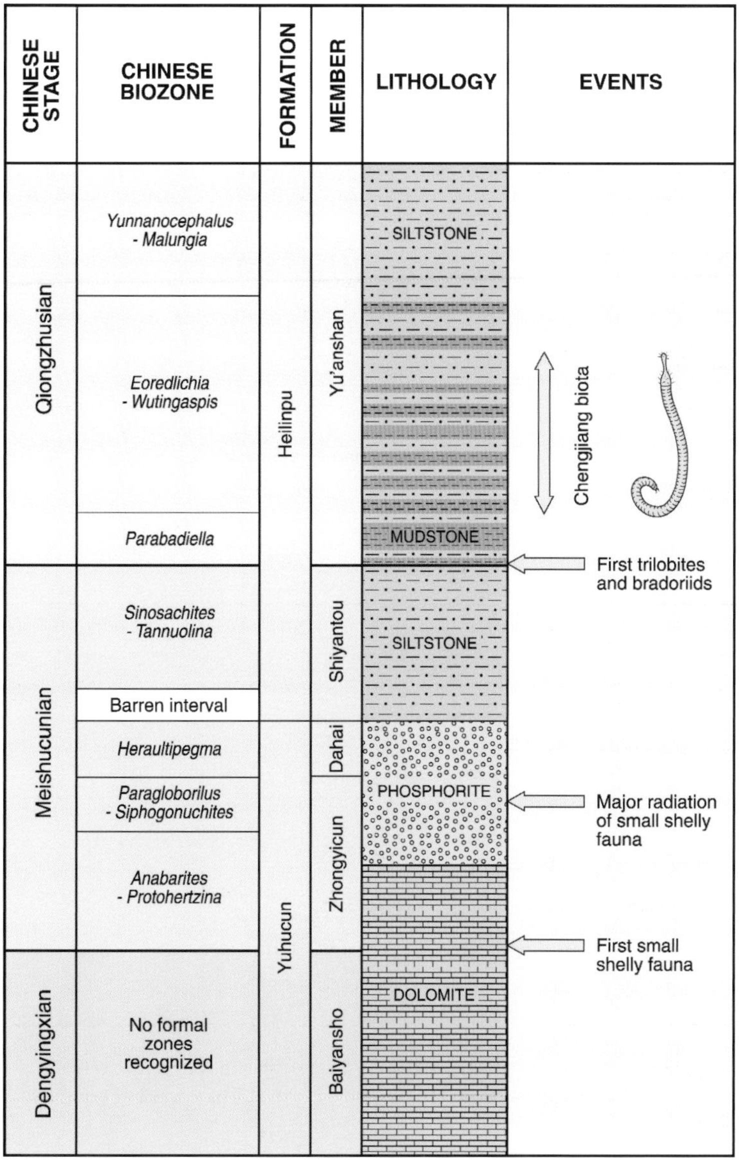

CHINESE STAGE	CHINESE BIOZONE	FORMATION	MEMBER	LITHOLOGY	EVENTS
Qiongzhusian	*Yunnanocephalus - Malungia*	Heilinpu	Yu'anshan	SILTSTONE	
	Eoredlichia - Wutingaspis				Chengjiang biota
	Parabadiella			MUDSTONE	First trilobites and bradoriids
Meishucunian	*Sinosachites - Tannuolina*		Shiyantou	SILTSTONE	
	Barren interval				
	Heraultipegma	Yuhucun	Dahai		
	Paragloborilus - Siphogonuchites		Zhongyicun	PHOSPHORITE	Major radiation of small shelly fauna
	Anabarites - Protohertzina				First small shelly fauna
Dengyingxian	No formal zones recognized		Baiyansho	DOLOMITE	

Figure 4.3 Stratigraphic occurrence of the Chengjiang biota and other faunal events within the Lower Cambrian of Chengjiang County, Yunnan Province (modified from data in Hou & Bergström 1997, Hou *et al.* 2002b). The Meishucunian (small shelly) biozones are as delimited in the nearby Meishucun and allied sections (Qian & Bengtson 1989, Qian *et al.* 2001).

and Dapotou has revealed a succession of four lithofacies, interpreted as representing a single transgressive–regressive sequence (Zhu *et al.* 2001a). Associated geochemical and basin analyses indicate that the depositional environment was influenced by storm-derived, fresh-water influxes from a (present day) westerly source area (Fig. 4.7). The oldest lithofacies (1), consisting of laminated siltstones with high organic carbon content, is thought to reflect deposition in shallow, oxygen-poor waters, early in the transgressive phase. The overlying, generally unbioturbated

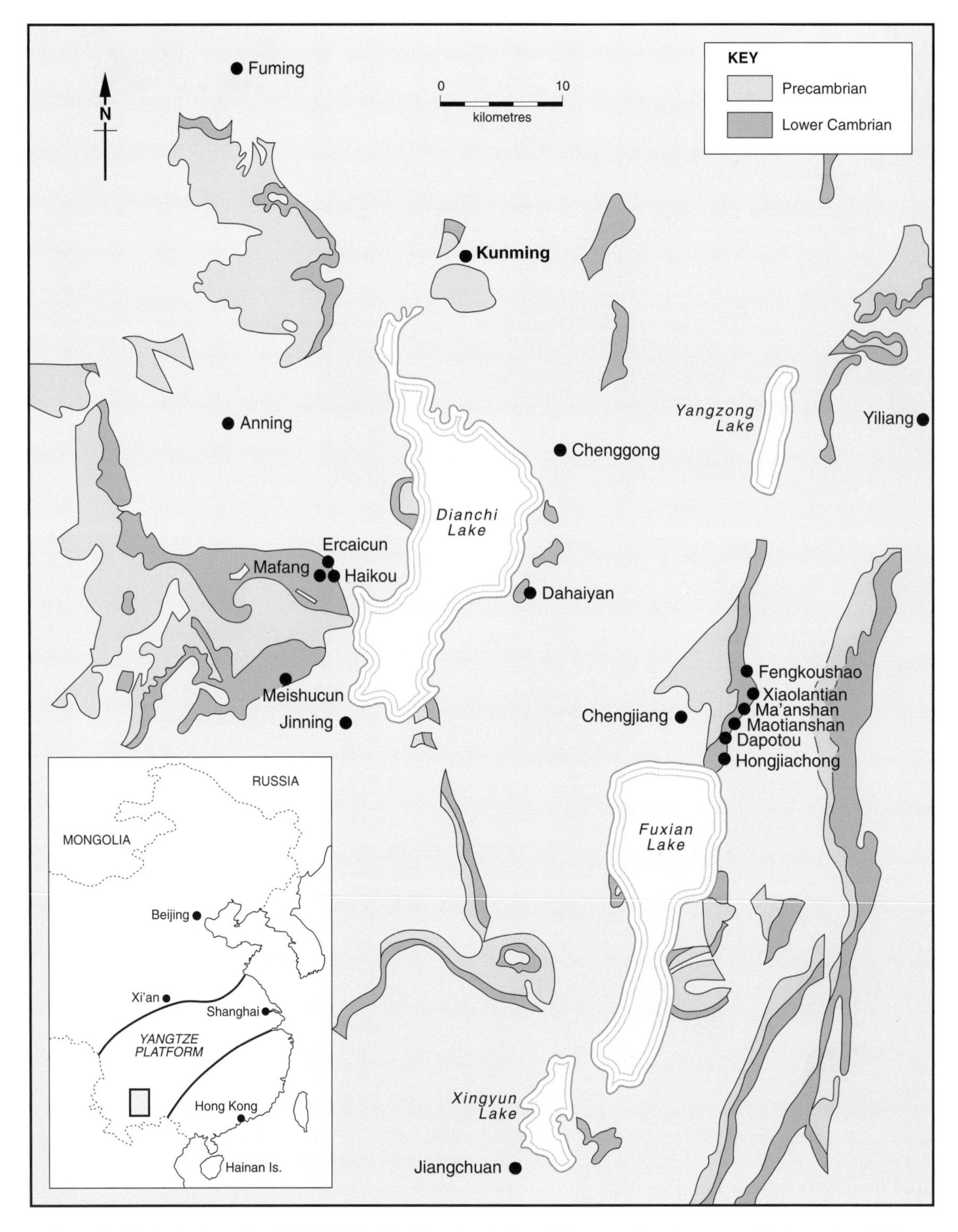

Figure 4.4 Geological map (scale 1 : 500,000) of the Kunming to Chengjiang area, showing some of the localities yielding the Chengjiang biota (after Zhu *et al.* 2001b).

Figure 4.5 Ercaicun, near Haikou, the site that yields the world's oldest known vertebrates.

Figure 4.6 Collecting at Mafang, near Haikou.

carbonaceous shales, with layers of pyrite and thin dolostone interbeds (lithofacies 2), are considered to have accumulated in a relatively deep, quiet-water, oxygen-poor and distal offshore setting that occasionally received influxes of silt from more proximal environments. Most of the Yu'anshan Member is formed of a lithofacies (3) that contains abundant soft-bodied fossils (essentially the "Maotianshan Shale" of earlier papers; see Ho 1942) and is modeled as a lower shoreface to proximal offshore environment of the early part of the regressive phase. The exceptionally preserved fossils occur in the upper, clay portion of "siltstone-clay" couplets whose components are interpreted as distal tempestite sediment together with muds deposited from suspension during fair-weather conditions. Host rocks to the soft-bodied fossils are apparently the so-called gray-type (algal-lacking) "claystones", which possibly formed by rapid deposition of suspended mud in waters that experienced frequent fresh-water inputs and were, therefore, below normal salinities. Depositional rates and a general lack of bioturbation are judged to be critical in helping to promote the exceptional preservation of the Chengjiang biota. In contrast, the co-occurring dark-gray, algal-bearing claystones are thought to reflect deposition in normal salinities, possibly in relatively dry seasons. The youngest of the four lithofacies (4), in the upper part of the Yu'anshan Member, consists of bioturbated silts and muds and is modeled as an upper shore face to nearshore environment.

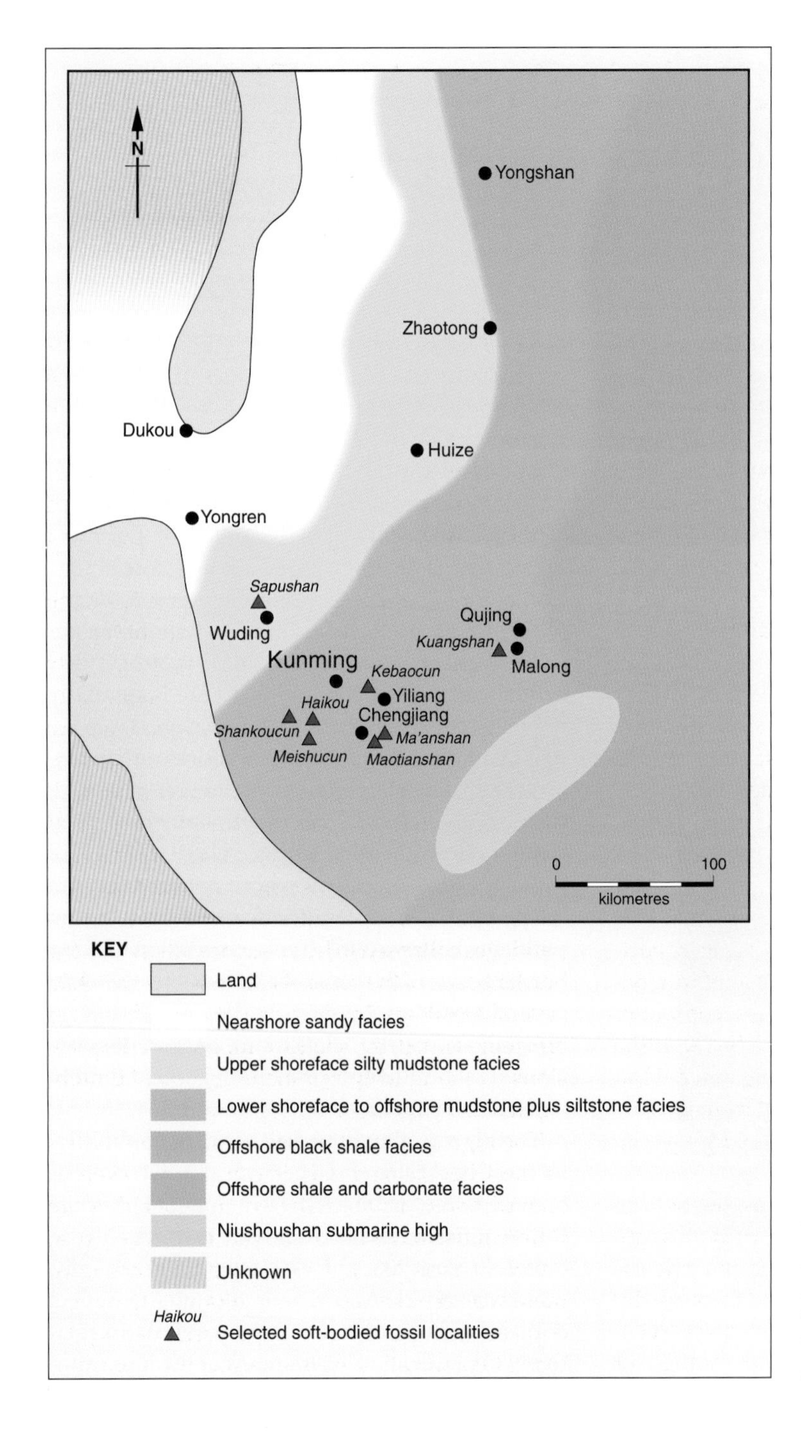

Figure 4.7 Paleogeographical and facies map of the eastern Yunnan area for the time of occurrence of the Chengjiang biota ("Maotianshan Shale", Yu'anshan Member; after Zhu *et al.* 2001a).

5 THE TAPHONOMY AND PRESERVATION OF THE CHENGJIANG FOSSILS

The sediment in which the Chengjiang fossils occur was deposited as soft mud, which has hardened to a yellow-weathering gray mudstone. The very fine grain of the rock has allowed the details of the fossils to be preserved with remarkable fidelity, although the processes that led to this preservation are currently poorly understood. Weathering has certainly played a part in the final appearance of the fossils, and calcium phosphate, for example, has been totally lost from the sediments and from the shells of animals such as the lingulate brachiopods.

Most of the fossils are in the form of flat, or nearly flat, impressions, although some retain a low three-dimensional relief. These impressions are sometimes dark in color, perhaps due to the presence of carbon, left behind as the original complex organic tissues degraded. It has been reported that the fossils are largely preserved by clay minerals (Zhu 1997, Babcock & Zhang 2001). Very commonly there is a reddish coloring imparted to the fossils by a thin rusty deposit of iron oxides; analyses of some of these show a content of up to 43% iron, in contrast with the 5–8% in the surrounding yellow mudstone (Bengtson & Hou 2001). The iron oxide seems to result from the oxidation of tiny dispersed framboids of iron pyrite (Chen & Erdtmann 1991).

The remains of hard tissues, such as the shells of brachiopods and the carapaces of trilobites, are well represented in the Chengjiang fauna, but less robust tissues, which are usually lost through decomposition, are also often beautifully preserved. These include the relatively tough cuticle of arthropods, which was probably similar to the outer skeleton of modern prawns, and truly soft tissues such as muscles, gills and intestines. These soft tissues would normally decay away very rapidly and disappear within a few weeks, or even days, of death. Many of the animals found in the Chengjiang fauna had no biomineralized skeleton at all and would not normally leave any trace in the fossil record.

Commonly, the carcasses show little evidence of decay and must, therefore, have been fossilized remarkably quickly after death, but in a number of specimens there are signs of some decomposition before final preservation. Babcock & Zhang (1997), for example, reported specimens of *Naraoia* in various states of disarticulation, from fully articulated with the appendages present, to articulated exoskeletons lacking appendages, to disjunct anterior and posterior shields; the completely disjunct specimens, however, may well include molted exoskeletons. Appendages of many of the arthropod fossils are visible through the carapace, where they are represented by a shallow furrow, apparently formed where the appendages collapsed to leave a space into which the harder tissue of the outer skeleton was pressed. In several fossil specimens some of the soft tissues are represented by shallow moulds or impressions in the muddy sediment and can be highlighted using incident light at a low angle.

The high quality of preservation of soft tissues in the Chengjiang muds is commonly taken as an indication that the sediment in which they are found must have been starved of oxygen (e.g. Babcock *et al.* 2001). This inhospitable environment would have kept scavengers away from the corpses, and would also have precluded widespread colonization of the sediment by burrowing animals, whose activity might otherwise have broken up the buried carcasses. Indeed, although trace fossils indicating the activity of burrowers do occur within the Chengjiang sediments (Zhu

1997), they are extremely rare in the horizons that yield the soft-bodied fossils (Hou *et al.* 1991). Some authors have suggested that the sediments may have been bound and sealed by microbial mats, which would have helped to inhibit decomposition of the carcasses (Chen & Erdtmann 1991).

There is little evidence of more than localized transport of the Chengjiang animals before they were entombed in the sediment. Indeed, some specimens clearly are found exactly where they lived. This is the case, for example, with some of the lingulid brachiopods, which are preserved with their shells lying flat on the bedding planes and their long pedicles extending obliquely down into the sediment below (Hou *et al.* 1991). It is likely that the animals were rapidly killed by asphyxia, although there is a difference of opinion as to whether this was caused by benthic sediment flows or by the incursion of anoxic waters (see Hagadorn 2002).

The features preserved vary between specimens of the same type of animal, and examination of several fossils is often required to reveal the full set of characters of an organism. The appearance of the fossils also varies with the orientation in which the dead animal rested on the sediment before it became compressed. This fact can be of value in reconstructing the living animals, because we can combine evidence from specimens compacted in lateral, oblique and dorsoventral attitudes to build up a three-dimensional picture of the organisms, rather in the way that we can envisage a building from plan and elevation diagrams. However, in the Chengjiang fossils, specimens appear almost always to have come to rest with their flattest surface parallel to the bedding, limiting the possibilities of applying this method of analysis. This fact also supports the conclusion that the carcasses suffered only gentle transport before being buried. The enclosing sediment must, however, have been deposited rapidly in order to bury and preserve the animals with minimal decomposition.

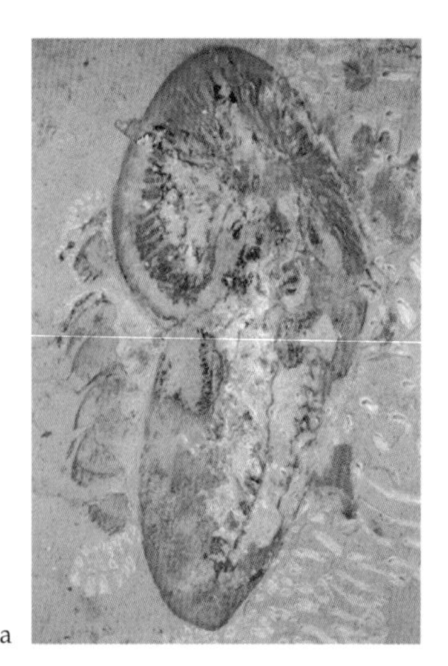

a

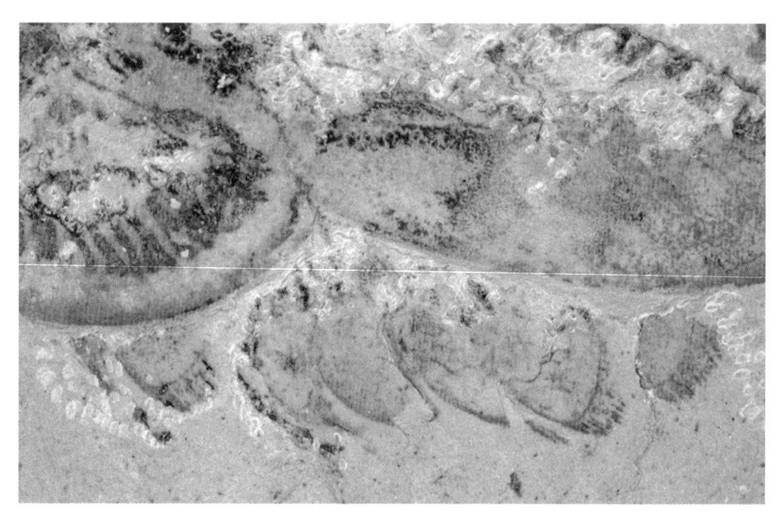

b

Figure 5.1 *Naraoia* sp. (a) Dorsolateral view of articulated specimen with appendages (RCCBYU 10278), ×2.7; Xiaolantian. (b) Detail of anterior exopods and lateral part of head and trunk shields (RCCBYU 10278), ×7.0.

6 THE PALEOECOLOGY OF THE CHENGJIANG BIOTA

Paleoecology is an inexact science. Paleontologists are naturally curious about the way that the animals, now found as fossils, originally lived their lives, but we have to be careful not to be drawn into untestable speculations. We can, however, utilize a number of lines of evidence and clues to help us to constrain our interpretations of the modes of life of past organisms. This is especially true where the animals are exceptionally preserved, as in the Chengjiang biota. In such circumstances we may, for instance, find evidence of the animals' food in their guts. Identifiable prey remains are occasionally apparent or, more commonly, we may find that the guts are mud-filled, indicating that the animal ingested sediment wholesale and then digested out the edible content of organic matter and micro-organisms.

Other clues may come from the nature of the enclosing sediment, which will reflect the environment in which the animal died. Details of the morphology of the animal also provide constraints; we can undertake rigorous functional morphological analyses of the hard and soft tissues of fossil organisms to determine how the animal was capable of moving and feeding. Comparisons with living relatives are also valuable, but for several of the Chengjiang animals there is little or nothing in the modern world to compare them to, so the help we can glean from extant analogies is somewhat limited.

Sedimentological studies of the Chengjiang mudrocks show that they comprise of sets of laminae that indicate rhythmic deposition or episodic sedimentary events. These events may have been turbidity flows (Chen & Zhou 1997) or storm-generated tempestites (Zhu *et al.* 2001a), which would have buried the organisms that were living on the seafloor. The muds that initially accumulated as these events waned may have been soft and soupy, and would have been difficult for new organisms to colonize. According to Chen & Zhou (1997), the pioneer community that began to re-establish itself in the muds consisted of burrowers (such as *Maotianshania*) and opportunistic taxa such as the bradoriids. As the substrate became more cohesive, an increasing diversity of burrowing (infaunal) and surface-dwelling (epifaunal) organisms moved in. Dead shells lying on the sediment may have provided sites for secure attachment of some individuals (e.g. *Archotuba* and the brachiopod *Longtancunella*).

Overall, the Chengjiang biota is remarkably diverse. Well over 100 animal species have been reported, referable to several major groups (Fig. 6.1). Counts show that arthropods account for more than 60% of the specimens found (Leslie *et al.* 1996), with "algae and bacteria" together comprising more than 30%; the combined specimen abundance of all the other taxa totals only about 3% of the biota. The majority of the animals were bottom-dwellers, with infaunal and epifaunal niches occupied by a variety of forms. Burrowers included the priapulid worms and the enigmatic *Facivermis*. Some of the lingulid brachiopods (*Lingulellotreta*, *Lingulella*) also probably lived in burrows or were supported above the sediment by partly buried pedicles. It has also been imaginatively suggested that they could have been opportunistically epiplanktonic, attaching themselves to floating organisms above the seafloor when the bottom conditions became inhospitable (Chen & Zhou 1997), but there is little evidence for this. The brachiopods would have been filter feeders, gaining their food from suspended organic matter in the water. Modern priapulids, in contrast, are

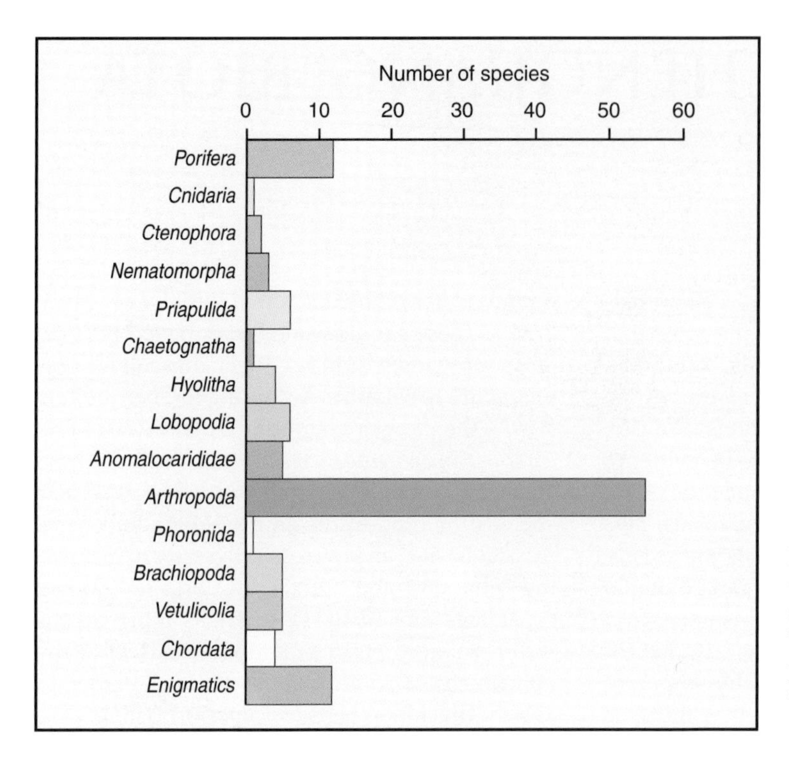

Figure 6.1 The composition of the Chengjiang fauna, based on the species list given at the end of this book. In addition to these groups, a possible annelid was illustrated by Chen *et al.* (1996) and Chen & Zhou (1997).

carnivores that use spines on their eversible proboscis to grasp prey.

Some of the epifaunal animals remained stationary (sessile), while others were more active and crawled about on the surface (vagile). The sessile forms were typically filter feeders, such as the sponges and the epifaunal lingulid brachiopods (*Longtancunella*, *Heliomedusa* and, probably, *Diandongia*). Among the more enigmatic sessile forms, the chancelloriid *Allonnia* may also have been a filter feeder, and the stemmed *Dinomischus* also probably gained its food from suspension in the water. These various filter feeders exploited different levels in the water column; *Heliomedusa* and *Diandongia* would have fed from the water just above the seabed, whereas *Longtancunella* was lifted above the seafloor by its pedicle. The sponges, chancelloriids and *Dinomischus* were raised even higher above the sediment and tapped levels above the brachiopods.

Many of the arthropods were members of the vagile epifauna, crawling across the sediment surface and maybe digging into the mud for food. The presence of mud-filled guts has been taken by most authors as evidence that many arthropods ingested the mud, processing out the organic content in their digestive systems (e.g. Bergström 2001). It has been suggested that these fills might be the mid-gut glands of the animals, three-dimensionally preserved initially by replacement with calcium phosphate and subsequently by clay minerals during weathering (Butterfield 2002). However, there is virtually no difference between the composition of the sediment-like gut fill and the host rock sediment, both consisting of the minerals quartz and muscovite (Hou *et al.* in press). Some of the arthropods, such as *Naraoia longicaudata*, were equipped with spinose gnathobases and would almost certainly have been capable of scavenging or of preying on small animals. Other

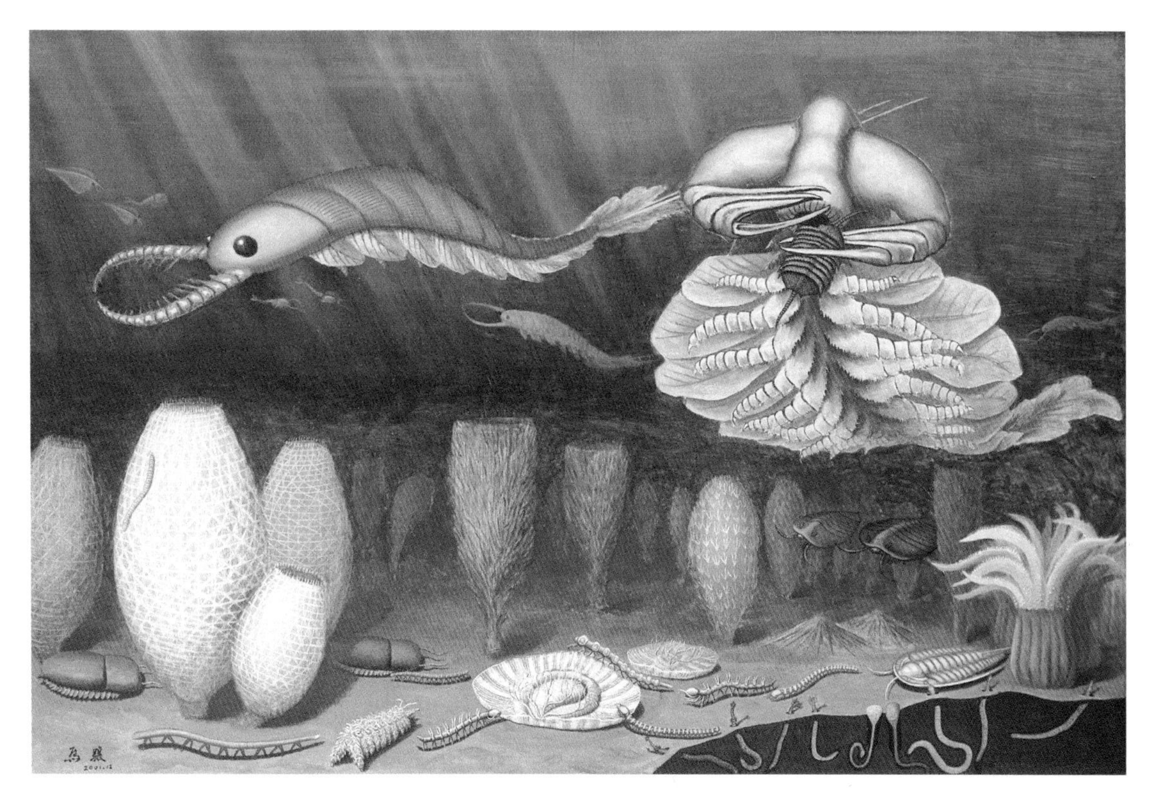

Figure 6.2 An artistic impression of the Chengjiang biota, by Ma Xiang of Kunming.

arthropods, for example *Waptia ovata*, appear to have organic remains in their guts, but no mud, indicating that they were selective in their feeding.

The lobopods were also mobile bottom dwellers. It has been suggested that they may have been predators and scavengers, feeding on sponges and other soft-bodied animals (e.g. Briggs *et al.* 1994).

The water above the seabed was colonized by a wide variety of floating and swimming animals. The floaters are represented by the comb jellies (ctenophores), and perhaps by the medusoid-like oddities, *Eldonia* and *Rotadiscus*. Dzik *et al.* (1997), however, have argued that the latter two were benthic epifaunal animals, lying on the seabed, where they became attachment sites for a number of epibionts.

Swimming organisms in the Chengjiang biota include the vertebrate *Myllokunmingia*, the anomalocarids, the vetulicolians, and a number of arthropods, including *Isoxys*. The anomalocarids were clearly fearsome predators, with large limbs at the front of the head to capture prey. Some arthropods, such as *Fortiforceps, Occacaris* and *Forfexicaris*, also possessed large appendages for grasping food.

Evidence for predation also comes from the occurrence of supposed fossil feces (coprolites) containing the remains of arthropod cuticle. Some of this cuticle probably comes from swimming arthropods, and anomalocaridids are likely producers of these coprolites. Other possible coprolites contain bradoriid valves, and these feces might well have been produced by the infaunal predatory priapulids as well as by larger arthropods (Chen & Vannier 2000, Babcock & Zhang 2001).

The Chengjiang biota clearly preserves a complex ecosystem with considerable niche diversification (Fig. 6.2). Trophic groups present include predators, scavengers, high and low level filterers and, it seems likely, deposit feeders. Apart from the coprolites and some epibionts, there is little evidence of the direct interactions that must have occurred between members of this community. However, it is evident that by this time in the Early Cambrian selection pressures had produced a structured biotic system with a diversity of life modes comparable to that seen in modern seas.

PART TWO CHENGJIANG FOSSILS

7 ALGAE

Algae are unicellular or multicellular organisms that undertake photosynthesis. A general difference from the photosynthetic mosses and tracheate plants is that they live in water. Photosynthesis is concentrated in colored organelles, the plastids, and there are three main groups of algae named after the particular color present—the red algae, the green algae, and the brown group. All of these algae are nucleate, eukaryotic organisms. Formerly a fourth group was included, the blue-green algae. These are now recognized as being anucleate (prokaryotes) and are commonly called cyanobacteria. The term "algae" does not have any phylogenetic meaning but rather indicates an algal mode of life.

Remains of both unicellular and multicellular algae have been found in the Precambrian, but it is often difficult to place them in particular algal groups. This is so with the four algal species, all thread-like apparently multicellular forms, that have been recorded from the Chengjiang Cambrian deposits: *Megaspirellus houi*, *Sinocylindra yunnanensis*, and *Yuknessia* sp. (all Chen & Erdtmann, 1991), and *Fuxianospira gyrata* Chen & Zhou, 1997.

Genus *Fuxianospira* Chen & Zhou, 1997

Fuxianospira gyrata Chen & Zhou, 1997

F. gyrata comprises an unbranched, cylindrical filament of uniform diameter that can reach 1.2 mm. It is tightly helicoidal along its length, and so when flattened has a beaded appearance. This species is the most abundant alga of those occurring at Chengjiang. Some specimens, at least, have been figured under the name *Yuknessia* (Chen *et al.* 1996, Hou *et al.* 1999). This genus was described originally by Walcott (1919) on the basis of his species *Yuknessia simplex* from the Middle Cambrian Burgess Shale, the macroscopic form of which resembles some modern tubular green algae (Conway Morris & Robison 1988).

Key References Chen & Erdtmann 1991, Chen *et al.* 1996, Chen & Zhou 1997.

Figure 7.1 *Fuxianospira gyrata*. A cluster of strands (RCCBYU 10211), ×2.5; Ercaicun.

Genus *Sinocylindra* Chen & Erdtmann, 1991

Sinocylindra yunnanensis Chen & Erdtmann, 1991

The filament of *S. yunnanensis* is about 0.3 mm wide and it may be least 20 mm long; it is unbranched, and its surface is smooth. The macroscopic features of this species were thought to resemble those of members of the blue-green Oscillatoriaceae (Chen & Erdtmann 1991), but it was also recognized that its dimensions are too large for bacterial affinities. The relatively loose coiling indicates a certain rigidity of the filament comparable to that of the modern laminariacean *Chorda*, and the lack of branching is also a similarity with this group of brown algae, indicating a possible relationship.

Key References Chen & Erdtmann 1991, Chen *et al.* 1996.

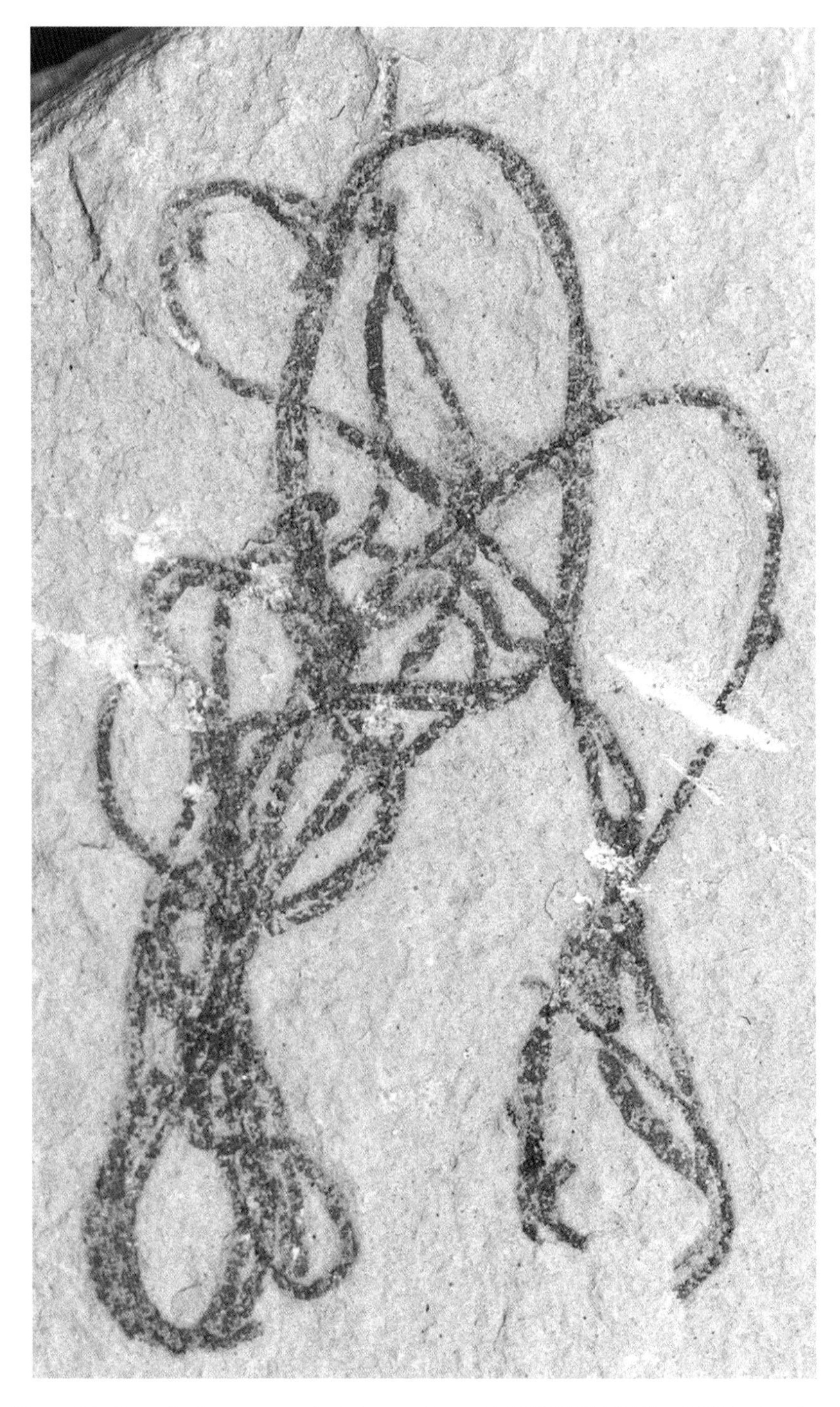

Figure 7.2 *Sinocylindra yunnanensis*. A cluster of strands (RCCBYU 10212), ×6.6; Maotianshan.

8 PHYLUM PORIFERA

Sponges are multicellular organisms that lack the sophisticated organization of true Metazoa. Organs and a nervous system are absent and the few types of cells that they have are not organized into tissues. The skeleton consists of a colloidal jelly, or of the horny material spongin, and/or a framework of siliceous or calcareous spicules that in some species is augmented with, or may consist entirely of, a calcareous skeleton. Their form varies from simple bag-shaped to stalked and elaborately branching colonies. In life they are sedentary and benthic. They filter feed from water currents pumped through minute holes (ostia) that lead to cavities lined with flagella-bearing cells (choanocytes) and thence to an interior cavity or water canal system (spongocoel or atrium); water is passed out via an apical opening (osculum). Most poriferans are shallow marine dwellers; there are also fresh-water and deep-water species. Today the latter include most of the hexactinellids, the glass sponges with their skeleton of opaline silica spicules. Sponge spicules and skeletons are common as fossils from the Cambrian onwards. This phylum is a diverse component of the Chengjiang biota; most of the described species are demosponges, a group in which the rays of the siliceous spicules usually diverge at 60° or 120°.

Genus *Triticispongia* Mehl & Reitner *in* Steiner *et al.*, 1993

Triticispongia diagonata Mehl & Reitner *in* Steiner *et al.*, 1993

T. diagonata is common within the Chengjiang biota, characteristically preserved as flattened limonitic and mouldic impressions. Specimens are small, some 6–10 mm high.

The sponge body is oval to rounded and thin walled, with spongocoel constituting most of the space. Some specimens display scattered basal root tuft spicules and spicule rays occur around the flattened oscular margin that appears as a more or less subhorizontal termination of the skeleton. The skeleton contains two moderately well organized series of small, delicate spicules with rectangular patterns. One series of spicules has its prominent spicule axes parallel and normal to the principal axis of the sponge, and the second series has a more diagonal arrangement.

This species is considered to be an early reticulosid hexactinellid sponge. Its root tuft presumably anchored the body to the substrate, as is characteristic of many Recent forms of glass sponge.

T. diagonata was originally described from the Cambrian of Hunan Province, China.

Key References Steiner *et al.* 1993, Rigby & Hou 1995, Hou *et al.* 1999.

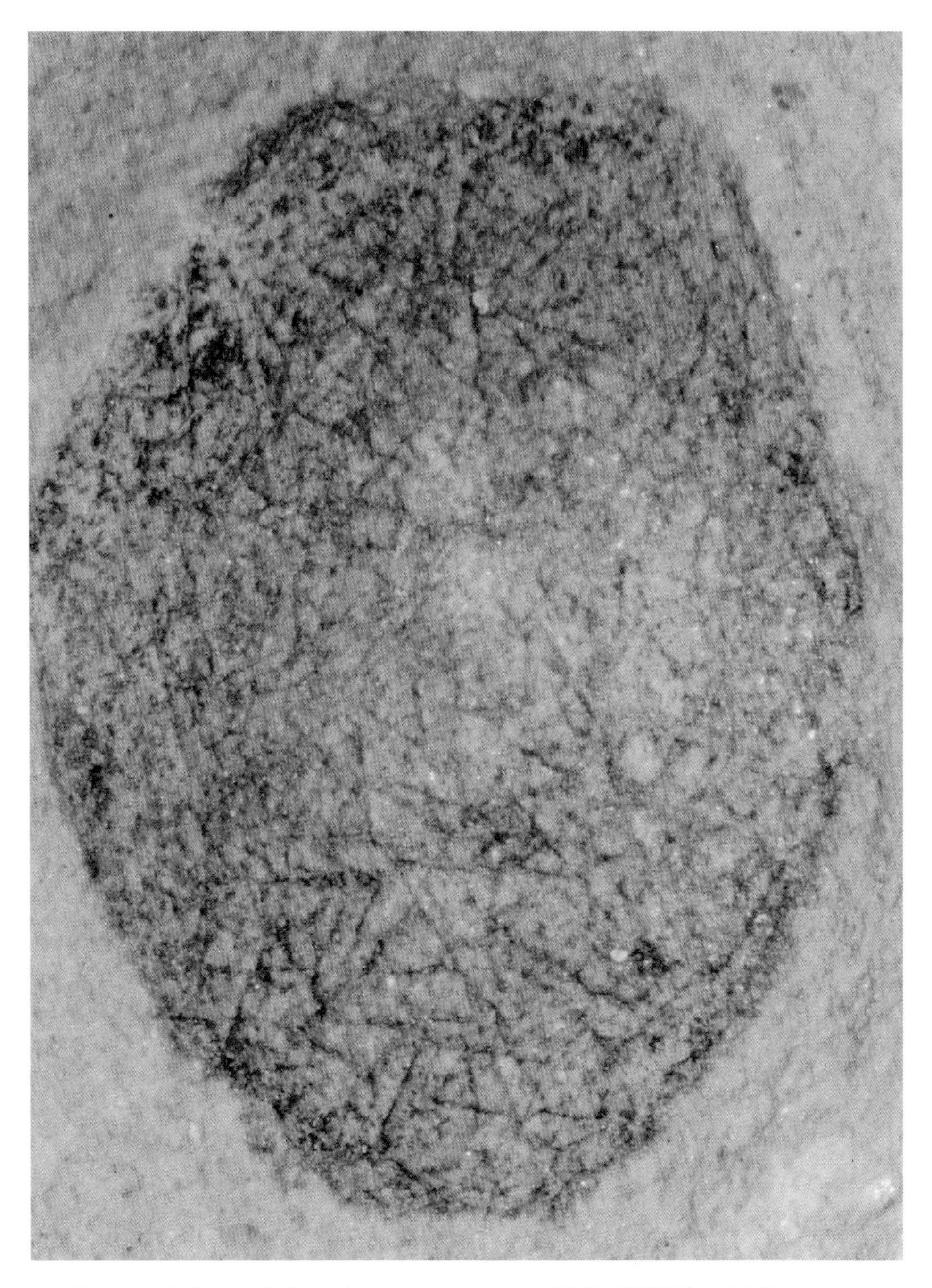

Figure 8.1 *Triticispongia diagonata*. A complete specimen (NIGPAS 115320), ×23.0; Xiaolantian.

Genus *Saetaspongia* Mehl & Reitner *in* Steiner *et al.*, 1993

Saetaspongia densa Mehl & Reitner *in* Steiner *et al.*, 1993

The generic name of this relatively common Chengjiang sponge alludes to its thin, hair-like spicules. Specimens from Yunnan Province are preserved as impressions, in low bas-relief.

This sponge has a moderately small, well-defined, almost circular body up to 3 cm by 4 cm in diameter. An oscular opening is not seen, though a flattened part of its outline may be the margin of the osculum. The skeleton consists mainly of very narrow diactine spicules, 1–2 mm long and about 0.025 mm in diameter. The spicules occur as dense, semiparallel, almost plumose bundles, and do not project beyond the outer margin of the sponge body. Less common, somewhat thicker spicules, up to 0.1 mm in diameter, are intermixed. The spicules may include three-, four-, or even six-rayed (hexactine) forms.

S. densa is the type species of a monotypic genus. The presence of small, probable hexactine spicules in specimens from Yunnan Province suggests that it may belong to the Hexactinellida, the glass sponges (Rigby & Hou 1995). Many Recent hexactinellids have a basal cluster of siliceous fibers for anchorage, but a root tuft is not evident in this species.

The original specimens of *S. densa*, from the Lower Cambrian Niutitang Formation of Hunnan Province, China, are generally larger than those from the Chengjiang biota.

Key References Steiner *et al.* 1993, Rigby & Hou 1995, Chen *et al.* 1996, Hou *et al.* 1999.

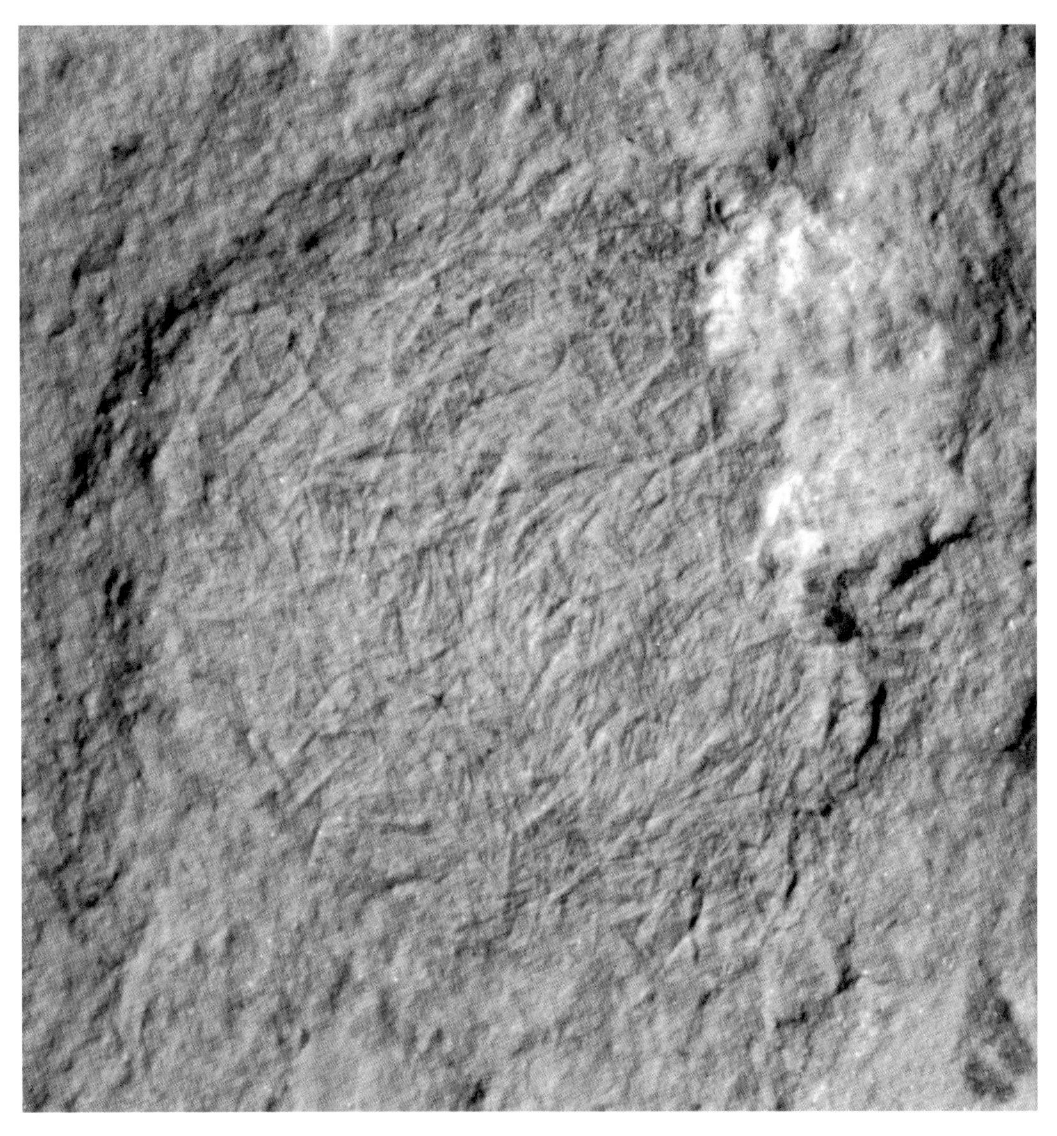

Figure 8.2 *Saetaspongia densa*. An almost complete specimen (NIGPAS 115321), ×14.0; Maotianshan.

Genus *Choiaella* Rigby & Hou, 1995

Choiaella radiata Rigby & Hou, 1995

C. radiata is relatively rare. The type material occurs as limonite replacements of thin, spicular, vertically and diagonally flattened skeletons.

In overall form the species is small and discoidal to low and broad shield-shaped or funnel-shaped. The skeleton consists essentially of a radiating thatch of small, apparently knobbly, single axis spicules (monaxons) that are generally of one size and which display bundling locally. Other than as a minor fringe, the spicules do not project beyond the margin of the disc.

The genus *Choiaella* is monotypic. It is assigned to the demosponges, the large class that includes most sponge species.

C. radiata is envisaged to be a three-dimensionally outwardly radiating tuft-like or brush-like sponge that eminates from a central point. Comparisons between this skeletal pattern and that of the Recent sponge *Radiella* suggest that *C. radiata* possibly had a similar, infaunal lifestyle (Rigby & Hou 1995).

This species is known only from localities in the Chengjiang area.

Key References Rigby & Hou 1995, Hou *et al.* 1999.

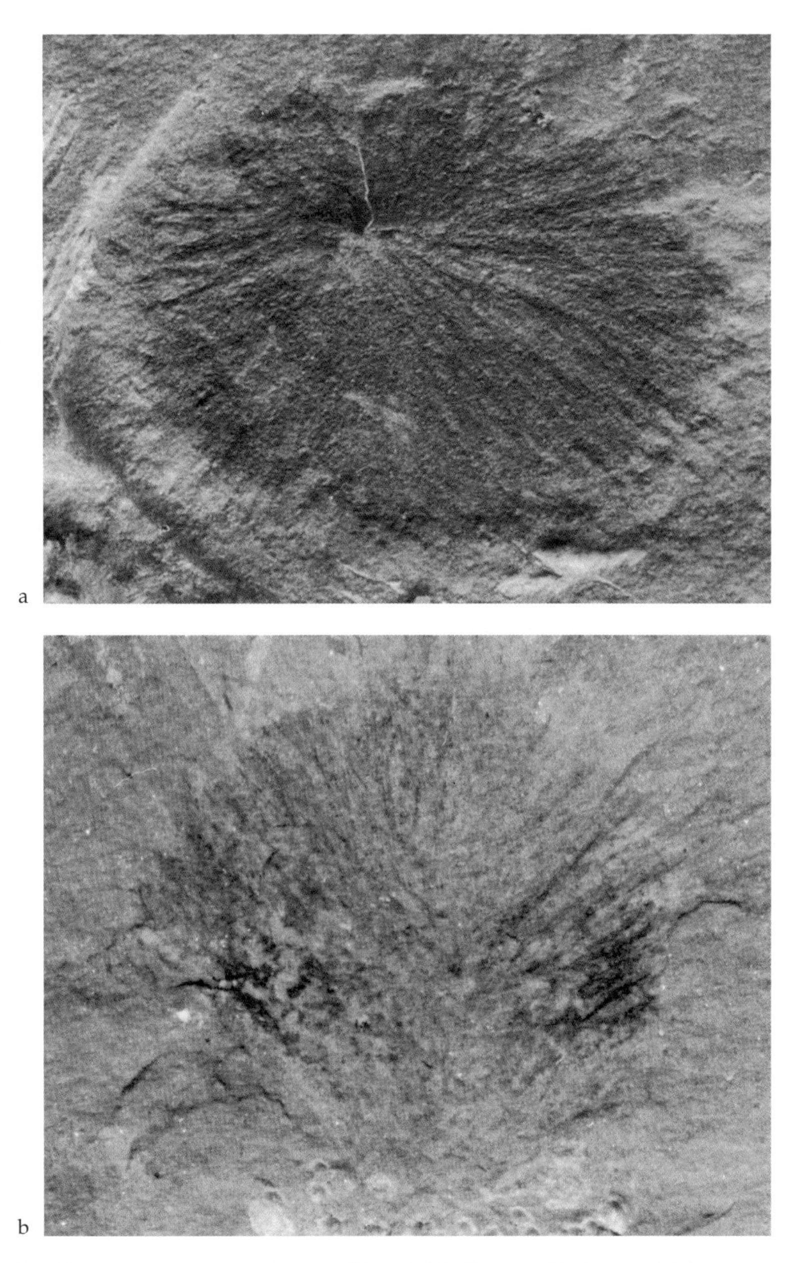

Figure 8.3 *Choiaella radiata*. (a) A complete specimen (NIGPAS 115324), ×13.1; Xiaolantian. (b) A complete specimen (NIGPAS 115328a), ×16.6; Maotianshan.

Genus *Choia* Walcott, 1920

Choia xiaolantianensis Hou, Bergström, Wang, Feng & Chen, 1999

This relatively common sponge occurs as small, attractively shaped impression fossils.

The body has a low, conical-shaped central disc up to 2 cm in diameter, composed of short, thin monaxon (single axis) spicules that combine to give a thatch-like appearance. Radiating from the disc there is a plethora of slender and discrete monaxon spicules, some of which are longer than the diameter of the disc itself.

Choia is the type genus of the demosponge Family Choiidae. The low, hat-like form, together with a lack of any obvious area of attachment, indicates that *Choia* may simply have rested on the substrate, convex side up, filtering food out of water that passed radially in through the disc and out via the central region (Rigby 1986). The Recent infaunal sponge *Radiella*, in which the skeleton is an upward and outward radiating structure, provides a possible model for *Choia*. Another reconstruction has *Choia* resting on the sediment with spicules radiating upward and outward in many directions from the disc, in pin-cushion fashion (Conway Morris 1998).

C. xiaolantianensis is known only from the Chengjiang biota. Congeneric species have been reported from the Middle Cambrian of British Columbia, Utah and Quebec (Rigby 1986).

Key References Rigby 1986, Rigby & Hou 1995, Hou *et al*. 1999.

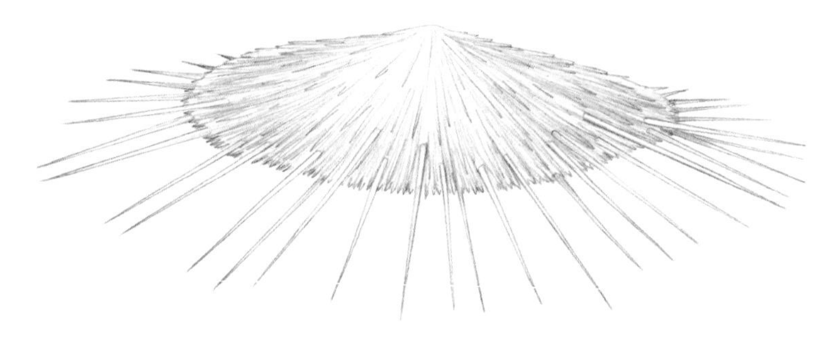

Figure 8.4 Reconstruction of *Choia* in inferred living position (after Rigby 1986).

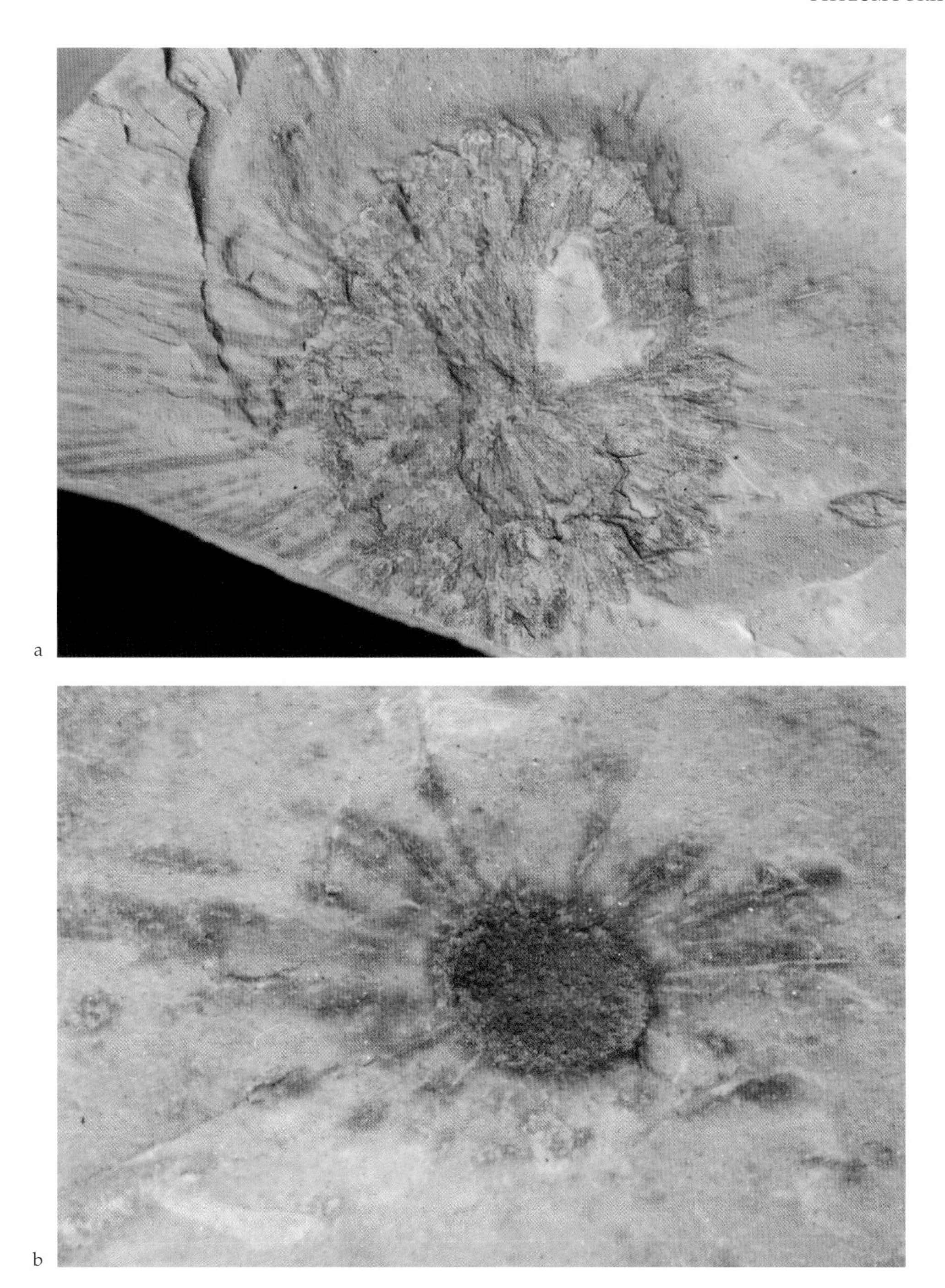

Figure 8.5 *Choia xiaolantianensis*. (a) A complete specimen (NIGPAS 115443), × 3.4; Xiaolantian. (b) A complete specimen (NIGPAS 115444), × 10.8; Xiaolantian.

Genus *Allantospongia* Rigby & Hou, 1995

Allantospongia mica Rigby & Hou, 1995

This is a rare species. The type material consists of fairly complete and fragmentary limonite impression fossils that have replaced thin spicular skeletons.

The body is relatively small (the holotype is 14 mm by 10 mm) and elongate ovate to sausage-shape in overall form. It consists mostly of a radiating thatch of small, single-axis (monaxon) spicules that are locally clumped into tufts. Some parts of the central area of the holotype skeleton are more open-textured. Moderately larger spicules extend outwards from around the central thatched area.

The demosponge genus *Allantospongia* is assigned to the Family Choiidae and is known only from a single species. The life orientation of *A. mica* is uncertain. The species is unknown outside the Chengjiang biota.

Key References Rigby & Hou 1995, Hou *et al.* 1999.

Figure 8.6 *Allantospongia mica*. A complete specimen (NIGPAS 115322), ×11.0; Xiaolantian.

Genus *Leptomitus* Walcott, 1886

Leptomitus teretiusculus Chen, Hou & Lu, 1989

L. teretiusculus is a moderately common, thin-walled sponge. Specimens range up to 110 mm long and about 12 mm wide.

The body is very elongate, tube-shaped and composed of two skeletal layers, each with single-axis (monaxon) spicules. The dominant, outer layer consists of a vertical thatch of fine and larger spicules, which in some specimens are slightly smaller in the lower half of the skeleton. The inner skeletal layer consists of tiny, more poorly defined, horizontally arranged spicules. A short fringe of spicules extends beyond the oscular margin.

Leptomitus is the type genus of the Family Leptomitidae and is generally recognized to represent the stock from which a variety of demosponges evolved (Rigby 1986). In life *L. teretiusculus* was attached to the substrate by a relatively small surface area, straining food from water pumped through its wall.

L. teretiusculus is known only from the Chengjiang biota. A larger *Leptomitus* species occurs in the Burgess Shale (Rigby 1986). Congeneric material has also been reported from the Lower and Middle Cambrian of Guizhou Province, China (Zhao *et al.* 1999a, Zhao *et al.* 1999c).

Key References Rigby 1986, Chen *et al.* 1989b, Rigby & Hou 1995.

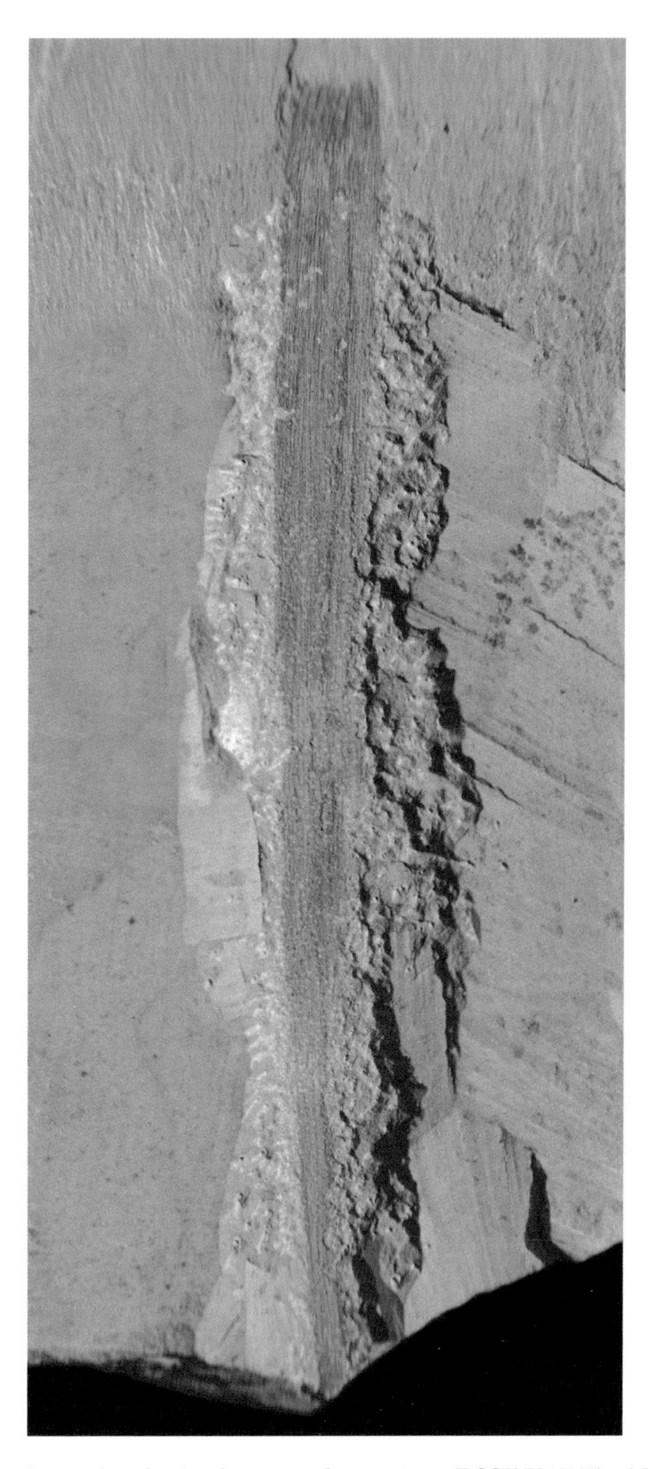

Figure 8.7 *Leptomitus teretiusculus*. An almost complete specimen (RCCBYU 10213), ×2.7; Maotianshan.

Genus *Leptomitella* Rigby, 1986

Leptomitella conica Chen, Hou & Lu, 1989

This is one of the more common sponge species in the Chengjiang biota. Specimens are flattened and have an overall conical shape a few centimeters long.

The sponge body is thin-walled. The skeleton is composed of single-axis (monaxon) spicules arranged in two layers. The outer layer consists of a vertical thatch of fine spicules and associated vertical rods. The inner layer is formed of both bundled and unbundled horizontally arranged spicules that are slightly finer than those in the outer layer.

L. conica belongs to the leptomitid family of demosponges. In life it was attached to the substrate at its narrow end. *Leptomitella* was originally described from the Middle Cambrian of Utah. *L. conica* is known only from the Lower Cambrian of Yunnan Province.

Key References Rigby 1986, Chen *et al.* 1989b, Chen *et al.* 1996, Hou *et al.* 1999.

Figure 8.8 *Leptomitella conica*. An almost complete specimen (NIGPAS 108485), with algae, ×8.7; Maotianshan.

Genus *Paraleptomitella* Chen, Hou & Lu, 1989

Paraleptomitella dictyodroma Chen, Hou & Lu, 1989

P. dictyodroma is a relatively common, tubular, thin-walled sponge that grew to about 10 cm in height. The fossils are flattened, and have a maximum width of 12 mm.

The base of the sponge is narrow, and the oscular margin seems to be rounded. Its double-layered skeleton is formed of single-axis (monaxon) spicules. The outer layer consists of coarse, slightly curved oxeas (monaxons having two rays, with both ends pointed) that interlock with one another to form tapering, elongate areas filled with fine, vertically arranged spicules. Bundles of horizontally arranged spicules make up the inner layer.

P. dictyodroma is the type species of the leptomitid genus *Paraleptomitella*, a demosponge that differs from the closely similar *Leptomitella* in the nature of the fabric of its outer skeletal layer. It is assumed that *P. dictyodroma* lived anchored to the substrate, but no evidence for a basal tuft structure has been documented.

The species has been found only in the Chengjiang biota.

Key References Chen *et al.* 1989b, Chen *et al.* 1996, Hou *et al.* 1999.

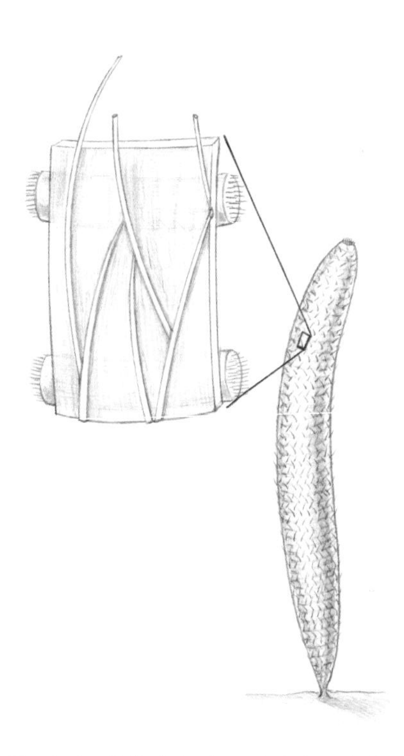

Figure 8.9 Reconstruction of *Paraleptomitella dictyodroma* in inferred life position, with detail of skeletal layers (after Chen *et al.* 1989).

Figure 8.10 *Paraleptomitella dictyodroma*. Two almost complete specimens (NIGPAS 108489, left; and NIGPAS 108490, right), ×1.8; Maotianshan.

Paraleptomitella globula Chen, Hou & Lu, 1989

This is a distinctively shaped, thin-walled sponge, with a maximum height of about 7 cm. The lower part is elongate and tubular, some 5 mm wide, above which it expands into a balloon-like shape of around 15 mm maximum width, with a much narrower osculum.

Both layers of the skeleton consist of single-axis (monaxon) spicules. The outer layer has an interweaving of slightly curved, coarse oxeas (monaxons having two rays, with both ends pointed), between which there is a network of fine, vertical spicules. The inner skeletal layer is comprised of fine horizontal spicules arranged in bundles.

P. globula differs from *P. dictyodroma*, the more commonly found type species, in the shape of the upper part of its body. It presumably lived attached to the seafloor, by its narrow end.

P. globula is confined to the Chengjiang biota.

Key References Chen *et al.* 1989b, Chen *et al.* 1996, Hou *et al.* 1999.

Figure 8.11 *Paraleptomitella globula*. An almost complete specimen (NIGPAS 108494), ×4.6; Maotianshan.

Genus *Quadrolaminiella* Chen, Hou & Li, 1990

Quadrolaminiella diagonalis Chen, Hou & Li, 1990

This large, elongate ellipsoidal sponge is known from a few tens of specimens. Its two-dimensional impression fossils are up to 30 cm long and about 12 cm wide, narrowing proximally and also distally towards the presumed site of the osculum. The skeleton consists of four layers of single-axis (monaxon) spicules, arranged into two nets each of two layers. The spicules of the outermost layer are coarse, relatively widely spaced and extend virtually the entire length of the sponge; those of the second layer are finer, more closely spaced and horizontal. The spicules of the two layers of the inner net trend diagonally in opposite directions.

Quadrolaminiella is the only known genus of the Family Quadrolaminiellidae. The genus was originally considered to be a demosponge, possibly derived from *Leptomitus*-like forms by the development of, *inter alia*, a thicker skeleton. *Quadrolaminiella* has also been compared with Lower Cambrian hexactinellids (Reitner & Mehl 1995).

The life position of *Q. diagonalis* was probably vertical, anchored to the substrate at its relatively narrow basal region. It is one of two *Quadrolaminiella* species, both of which are found only in the Chengjiang biota.

Key References Chen *et al*. 1990, Reitner & Mehl 1995, Chen *et al*. 1996.

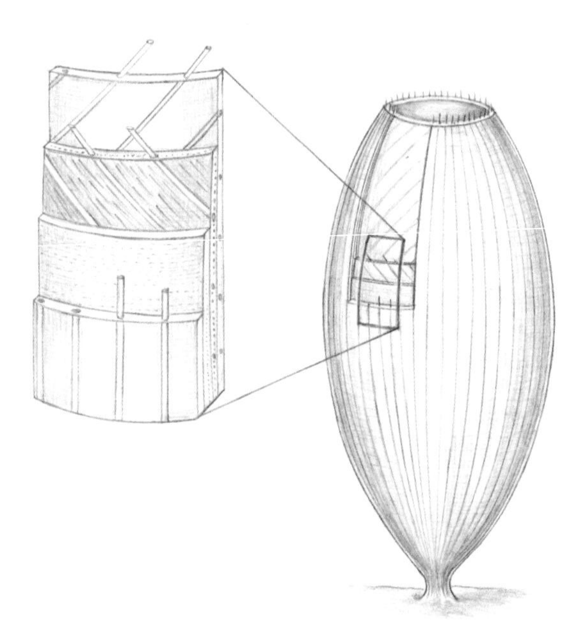

Figure 8.12 Reconstruction of *Quadrolaminiella diagonalis* in inferred life position, with a detail of the skeletal layers (after Chen *et al*. 1990).

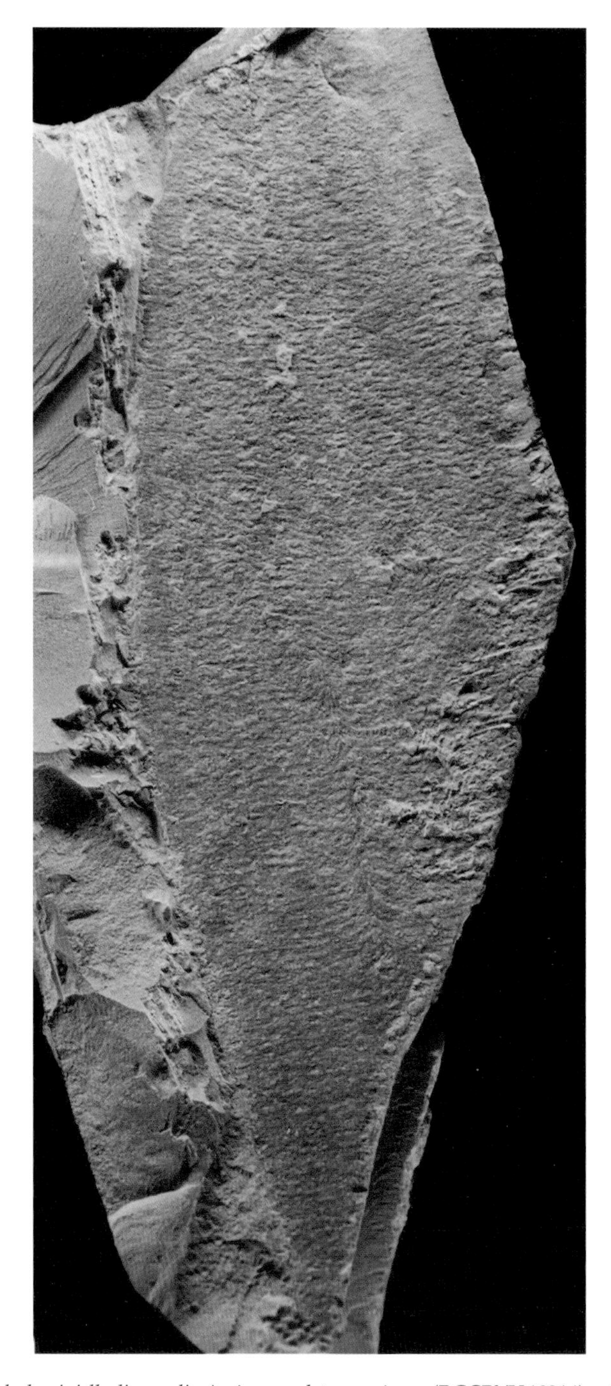

Figure 8.13 *Quadrolaminiella diagonalis*. An incomplete specimen (RCCBYU 10214), ×1.1; Maotianshan.

9 PHYLUM CNIDARIA

Cnidarians are primitive solitary and colonial metazoans represented by the corals, sea anemones and sea pens (anthozoans), jellyfish (scyphozoans), and hydrozoans (e.g. *Hydra*). They have radial symmetry, a body wall of only two layers (endoderm and ectoderm) and a body cavity (enteron) leading to an opening that serves as a mouth and an orifice for waste disposal. Thus, these diploblastic metazoans differ fundamentally in body plan from other metazoans, which are characterized by having three body wall layers (triploblastic) and bilateral symmetry. Most cnidarians are marine dwellers. A pelagic larval stage (the medusa) is common, but as adults most cnidarians are benthic and anchored, with the polyps collecting food with tentacles that ring the mouth. Only one or two purported species of the phylum are known from Chengjiang assemblages, including a questionable pennatulacean (Zhang & Babcock 2001).

Genus *Xianguangia* Chen & Erdtmann, 1991

Xianguangia sinica Chen & Erdtmann, 1991

This Chengjiang species, known from only a few tens of specimens, consists entirely of the remains of soft tissues. Most specimens are preserved laterally collapsed; a few are compressed dorsoventrally, giving an overall flower-like appearance to the fossil.

The animal appears to be approximately cylindrical, is typically about 60 mm high and has a maximum width of 30 mm at its base. It consists of a seemingly smooth pedal disc, a column with external strip-like longitudinal structures, extending into a series of about 16 strip-like or tentacular structures forming a flexible ring around what is assumed to be a centrally positioned mouth. The pedal disc is separated from the column by a conspicuous constriction. Each tentacle appears to taper gradually to a point.

This species has generally been allied to the sea anemones (Chen & Erdtmann 1991); the longitudinal structures of the column may correspond to internal folds (mesenteries) of the gut. Alternatively, but perhaps less likely, the strip-like structures might be the comb rows of a ctenophore ("comb jelly"; see *Maotianoascus octonarius*), a group commonly considered to be closely related to the cnidarians. Both ctenophores and a possible sea anemone, Walcott's (1911b) *Mackenzia costalis*, have been identified from the Burgess Shale.

X. sinica is known only from the Lower Cambrian of Yunnan Province.

Key References Chen & Erdtmann 1991, Chen *et al.* 1996, Hou *et al.* 1999.

Figure 9.1 *Xianguangia sinica*. (a) Lateral view (NIGPAS 108506a), ×2.6; Maotianshan. (b) Lateral view (RCCBYU 10215), ×1.9; Maotianshan. (c) Dorsal view (RCCBYU 10216), ×4.8; Maotianshan.

Figure 9.2 Reconstruction of *Xianguangia sinica* (after Chen & Erdtmann 1991).

10 PHYLUM CTENOPHORA

The ctenophores, otherwise referred to as comb jellies, sea combs, sea walnuts or sea gooseberries, are a minor group of Cambrian to Recent pelagic carnivorous metazoans. All Recent forms, some 90 species, are marine. They have a delicate, transparent, gelatinous and luminescent bag-like body that bears longitudinally arranged bands (comb rows), each with transverse lines of cilia (combs) throughout. At one end of the typically spherical body there is a centrally positioned mouth. Many species have a pair of long, branched, contractile tentacles. Individuals are propelled through the water by the co-ordinated action of the cilia. Food is captured either by direct engulfment into the mouth, or by first entrapping the prey on the sticky tentacles. Some schemes have placed ctenophores close to deuterostomes (Nielsen 1995, 1998), but traditionally and commonly they are considered to be closely related to the Cnidaria (Nielsen 2001). Ctenophores are triplobasts and although not bilateral animals they do have some cell and tissue characteristics akin to those of the Bilateria. The monospecific *Maotianoascus* and *Sinoascus* (both Chen & Zhou 1997) are the only ctenophores known from the Chengjiang biota.

Genus *Maotianoascus* Chen & Zhou, 1997

Maotianoascus octonarius Chen & Zhou, 1997

This entirely soft-bodied animal has been described from just a few specimens. The fossils are preserved in two dimensions in lateral aspect.

The body is, overall, spherical in shape in lateral view. It is formed of eight longitudinally extending petaloid lobes, which have a ring-like arrangement that implies radial symmetry. Each of the lobes has a medial fold, along each face of which there is a longitudinal strip of fine, transversely arranged ridge and furrow-like structures flanked by smooth longitudinal areas. These delicate features are interpreted as the combs of comb rows. The lobes converge toward each other and meet in a small button-like dome at what is presumed to be the aboral end. In the holotype, the lobes appear to extend into a short, delicate, skirt-like membrane around the presumed site of a wide, centrally positioned mouth.

This species bears a compelling resemblance to Recent ctenophores. Cambrian ctenophores apparently have many more than the eight comb rows characteristic of their Recent counterparts. There are 16 rows in *M. octonarius* and 24 rows in the Burgess Shale ctenophore *Ctenorhabdotus capulus* Conway Morris & Collins, 1996. The supposed Chengjiang cnidarian *Xianguangia sinica*, also has longitudinally arranged strip-like (tentacular) structures that are 16 in number.

Locomotion in the pelagic *M. octonarius* was presumably achieved by the action of the multitude of tiny cilia on each comb row. *C. capulus* apparently lacked tentacles, and therefore it is likely to have captured its food directly into its mouth (see Conway Morris 1998). *M. octonarius* may have had a similar mode of feeding.

M. octonarius has been found only in the Lower Cambrian of Yunnan Province.

Key References Chen *et al.* 1996, Chen & Zhou 1997, Hou *et al.* 1999.

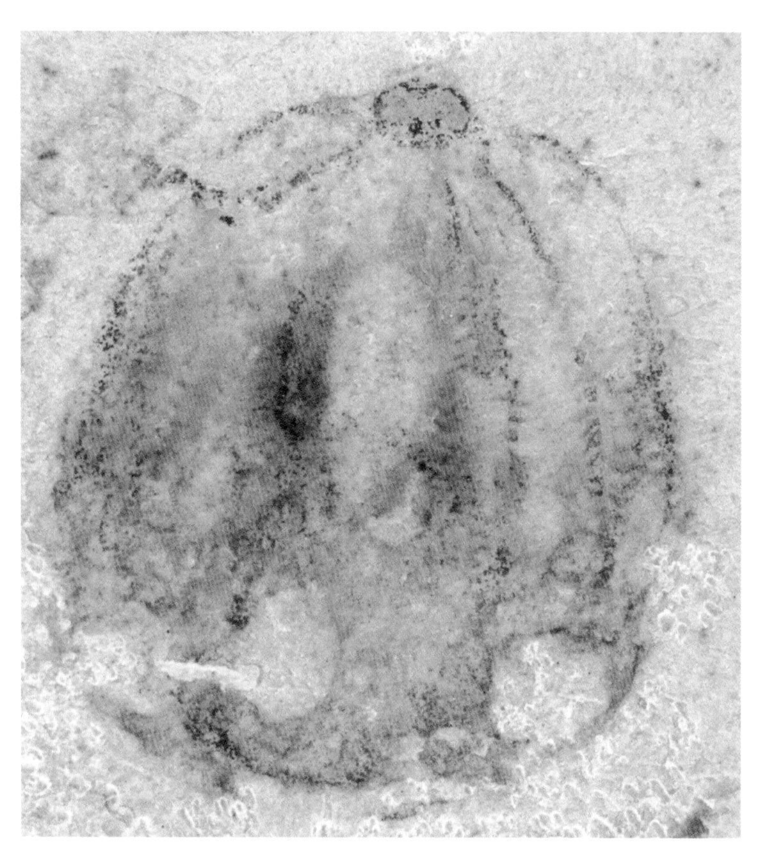

Figure 10.1 *Maotianoascus octonarius*. Oblique lateral view (RCCBYU 10217), ×14.5; Maotianshan.

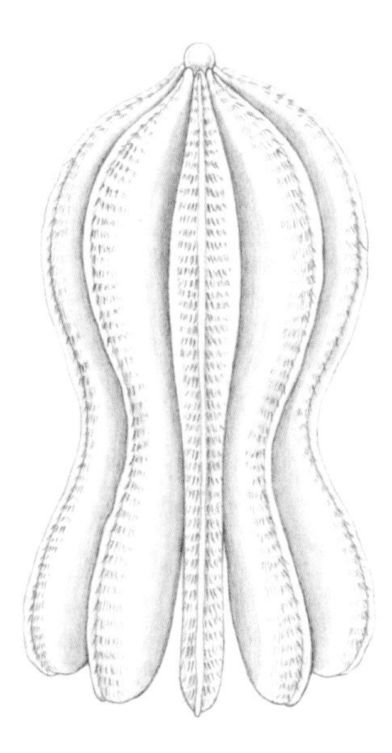

Figure 10.2 Reconstruction of *Maotianoascus octonarius*.

11 PHYLUM NEMATOMORPHA

This minor group, generally known as horsehair worms, is represented by about 300 living species (Ruppert & Barnes 1996). Except for the pelagic swimmer *Nectonema*, which lives in marine waters, all Recent nematomorphs inhabit fresh water. Most adult nematomorphs are 5–10 cm long; they have elongate hair-like bodies, a poorly differentiated head region and are free-living. In contrast, the larval stage, which has a protrusible and spine-bearing proboscis, lives parasitically and even molts within the body of a wide variety of hosts such as centipedes, grasshoppers, beetles, crabs, shrimps and other animals that live in or near water. Some authors recognize three possible nematomorph genera from the Chengjiang biota (Hou & Bergström 1994); other investigators reject a nematomorph assignment for some or all of these genera (e.g. Chen & Zhou 1997).

Genus *Cricocosmia* Hou & Sun, 1988

Cricocosmia jinningensis Hou & Sun, 1988

This worm is known from thousands of compression fossils. In some cases its sclerites are preserved in slight relief. The reddish colored preservation that is typical of the fossils contrasts with the generally lighter, buff hue of the host rock.

Large specimens are about 50 mm long and 2.5 mm wide. Anteriorly, there is a relatively long proboscis, the surface of which is armed with spines. The anterior-most region of the trunk displays narrow, unornamented annuli (about 40 per cm), whereas the annuli of the main trunk region are wider (20–24 per cm) and each bears a pair of cone-shaped sclerites arranged in longitudinal rows. The straight, simple gut stands out as a dark band extending from the proboscis to the posterior termination of the trunk, projecting from which there is a small curved spine.

C. jinningensis was originally considered, not least because of its proboscis morphology, to be a priapulid worm, a view maintained by some authors. Hou & Bergström (1994) subsequently allied the species, together with *Palaeoscolex* and *Maotianshania* from Chengjiang, to the palaeoscolecid worms, a group that they thought had some morphological similarities to priapulids, nematodes and especially nematomorphs, although they were not parasitic.

The fact that many specimens of this worm have a gut filled with mud suggests that it may have been an infaunal deposit feeder.

C. jinningensis is known only from the Lower Cambrian of Yunnan Province. Zhao *et al.* (1999b) have identified possible *Cricocosmia* from the Middle Cambrian Kaili biota of Guizhou Province, China.

Key References Hou & Sun 1988, Hou & Bergström 1994, Chen *et al.* 1996, Hou *et al.* 1999.

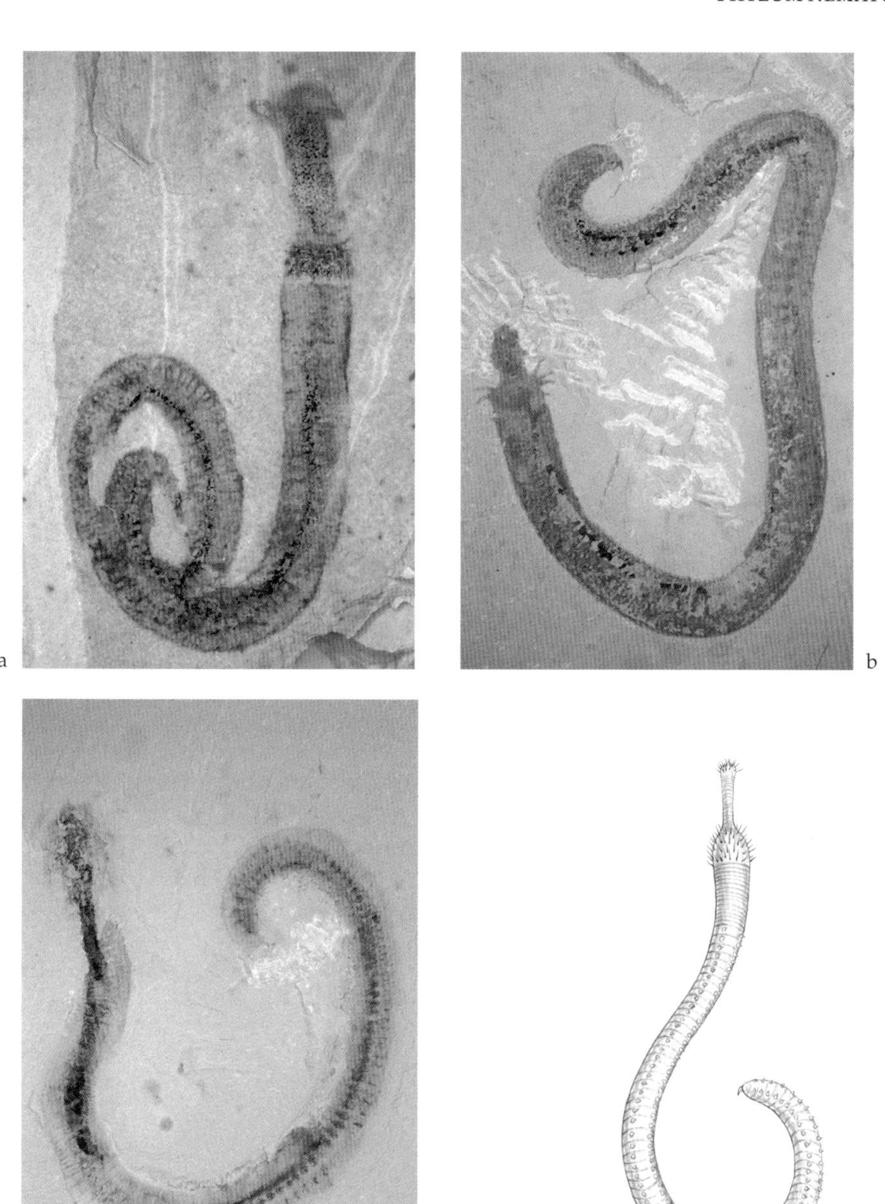

Figure 11.1 *Cricocosmia jinningensis*. (a) Lateral view of a complete specimen (RCCBYU 10219), ×4.2; Mafang. (b) Lateral view of a complete specimen (RCCBYU 10220), ×3.2; Mafang. (c) Lateral view of a complete specimen (RCCBYU 10221), ×3.6; Mafang. This species is also illustrated in the frontispiece.

Figure 11.2 Reconstruction of *Cricocosmia jinningensis*.

Genus *Palaeoscolex* Whittard, 1953

Palaeoscolex sinensis Hou & Sun, 1988

This relatively large worm is a common element of the Chengjiang biota. It occurs as complete and fragmentary compression fossils, consisting of soft parts and embedded biomineralized elements. Many specimens are coiled.

P. sinensis has a proboscis anteriorly, a spine posteriorly, and a trunk region up to 100 mm long and 4 mm wide comprising about 50 annuli per centimeter. Externally each annulus bears a double transverse row of tiny (30–45 μm diameter), approximately rounded plates (40–50 per annulus), in between which there is a ridge bearing a double transverse row of smaller, more numerous pits (60–80 per annulus). The gut is preserved as a fairly straight, dark band.

Palaeoscolex was originally considered to be an annelid (Whittard 1953), but has since been allied to, amongst others, the nematomorph worms (Hou & Bergström 1994), the priapulid worms (Chen & Zhou 1997, Conway Morris 1997a) or treated under phylum uncertain (Müller & Hinz-Schallreuter 1993). Isolated palaeoscolecidan sclerites may be represented in the fossil record by the hadimopanellids and like forms, a range of small, button-like phosphatic Cambrian microfossils. According to one notion, palaeoscolecids may in essence be armored priapulids and possibly provide a link between that group and the arthropods and other phyla that molt their cuticle (see Conway Morris 2000).

Most palaeoscolecidans may have been infaunal. Their plates possibly provided protection from sediment abrasion and predators and may have acted as aids in possible burrowing activities. However, the gut in *P. sinensis* seemingly lacks mud infilling, suggesting a non-deposit feeding, possibly epifaunal and predatorial lifestyle.

P. sinensis is unknown outside the Lower Cambrian of Yunnan Province. Congeneric material has been reported from the Middle Cambrian Kaili biota of Guizhou Province (Zhao *et al.* 1999a, 1999b) and from deposits as young as the Silurian outside China.

Key References Whittard 1953, Hou & Sun 1988, Müller & Hinz-Schallreuter 1993, Hou & Bergström 1994, Chen *et al.* 1996, Conway Morris 1997a, Hou *et al.* 1999.

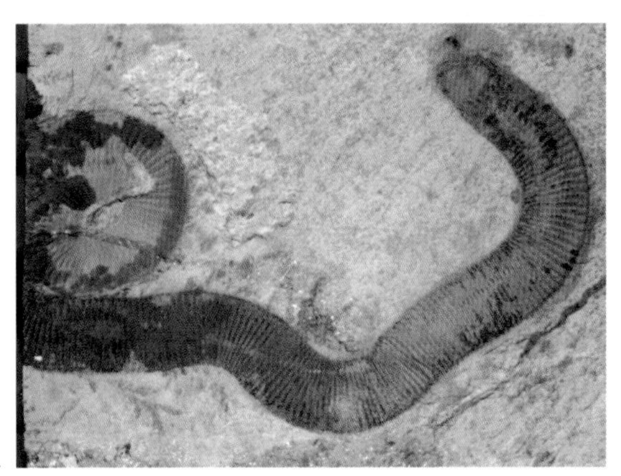

a

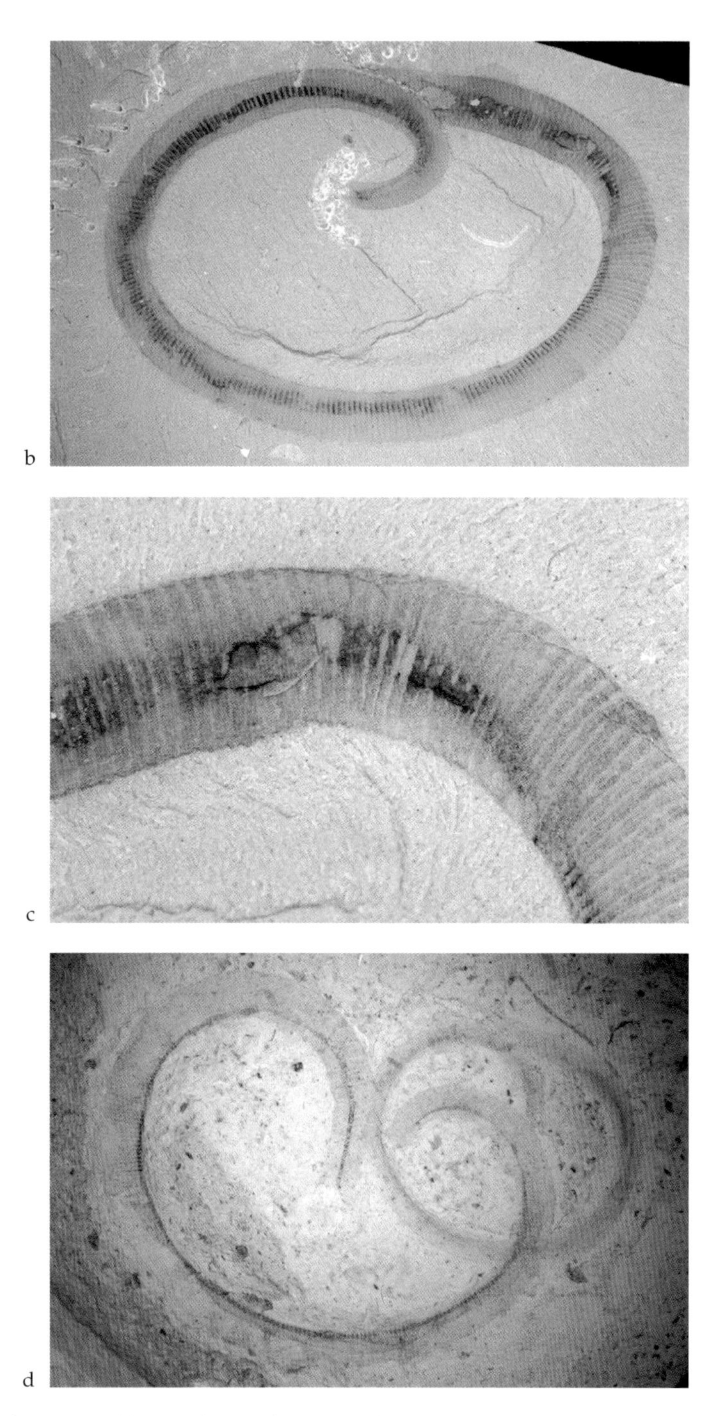

Figure 11.3 *Palaeoscolex sinensis*. (a) Lateral view of an incomplete specimen (RCCBYU 10222), ×2.9; Ma'anshan. (b) Lateral view of a complete specimen (RCCBYU 10223), ×4.0; Ercaicun. (c) Detail of the trunk (RCCBYU 10223), ×11.2. (d) Lateral view of two complete specimens (RCCBYU 10224), ×2.3; Ercaicun.

Genus *Maotianshania* Sun & Hou, 1987

Maotianshania cylindrica Sun & Hou, 1987

M. cylindrica is known from thousands of specimens, many of which show fine details of soft parts. Specimens occur as flattened body fossils, typically parallel to bedding. The head and most of the trunk is normally fairly straight to gently curved, whereas in many specimens the posterior region is preserved coiled through more than 300°.

The largest individuals are about 40 mm long and 2 mm wide. The relatively slender trunk has slightly convex, narrow annuli (3–4 per mm), each covered with many tiny pits that are assumed to represent the site of sclerites. The furrows delimiting the annuli have similar irregularly distributed pits. The anterior retractable proboscis bears many papillae and rings of neatly arranged spines. The bluntly rounded posterior normally terminates in a pair of tiny curved hooks. Similar structures are found in the Chengjiang worms *Cricocosmia jinningensis* and *Palaeoscolex sinensis*. The intestine is long, narrow and straight and is preserved as a dark film, with slight relief in some specimens.

Some authors consider *Maotianshania* to be a palaeoscolecid and ally that group to nematomorph worms (e.g. Hou & Bergström 1994). Others have highlighted the presence of its priapulid-like features (Sun & Hou 1987b, Chen & Zhou 1997, Conway Morris 1997a). Many specimens show a mud-filled intestine, indicating that *M. cylindrica* was probably a deposit feeder. Simple tubular burrows and trails in the host mudstone probably attest to the activities of various shelf-dwelling worms.

M. cylindrica occurs in the Lower Cambrian of Yunnan Province, especially the Chengjiang region. The genus has also been reported from the Middle Cambrian Kaili Lagerstätte of Guizhou Province (Zhao *et al.* 1999a, 1999b).

Key References Sun & Hou 1987b, Hou & Bergström 1994, Chen *et al.* 1996, Hou *et al.* 1999.

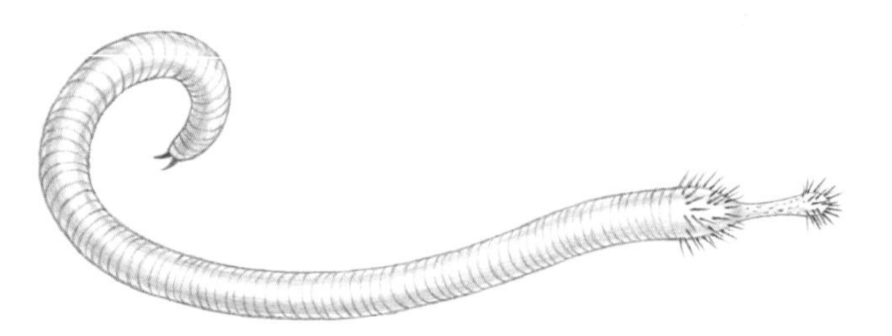

Figure 11.4 Reconstruction of *Maotianshania cylindrica*.

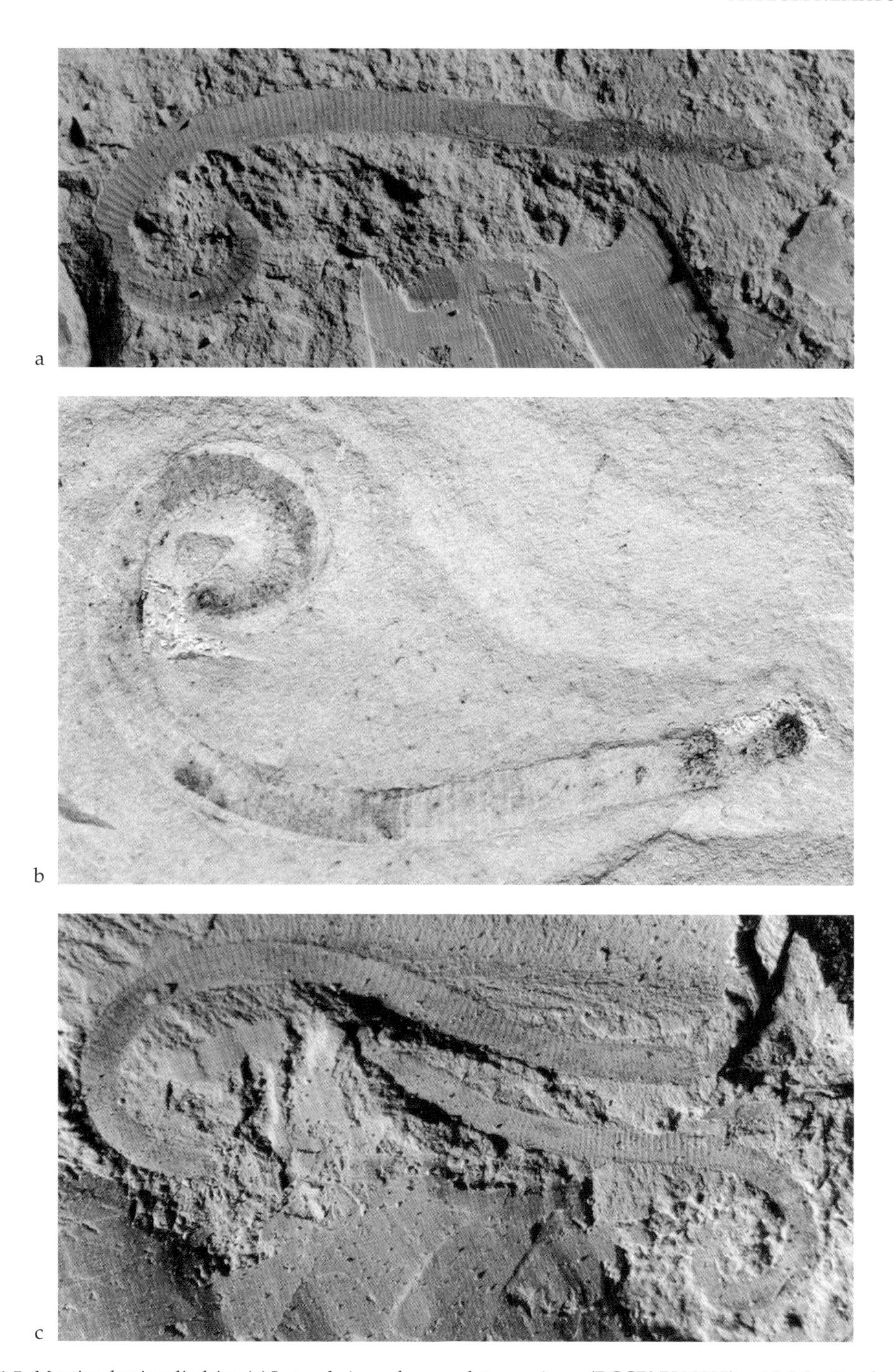

Figure 11.5 *Maotianshania cylindrica*. (a) Lateral view of a complete specimen (RCCBYU 10225), ×4.3; Maotianshan. (b) Lateral view of a complete specimen (RCCBYU 10226), ×4.8; Jianbaobaoshan, near Dapotou. (c) Lateral view of a complete specimen (RCCBYU 10227), ×4.7; Maotianshan.

12 PHYLUM PRIAPULIDA

The Chengjiang biota contains several species of priapulids, a group of non-segmented coelomate worms. The Recent species, which range to over 30 cm in length, live buried in shallow to deep marine sands and muds as carnivores and possibly deposit-feeding bacteriovores. Their cylindrical body consists of a trunk and, anteriorly, a shorter barrel-shaped proboscis that can invaginate into the anterior part of the trunk and is used in feeding and burrowing activities. The proboscis bears a terminal mouth and has many tiny cone-shaped external projections (scalids) arranged in transverse or longitudinal rows. In some species the trunk bears tiny spines and also posterior extensions that are used to anchor the worm to the substrate. Based on molecular sequence and other data, priapulids have been allied with the arthropods and nematodes in a major grouping of triploblastic animals that molt their cuticle, called ecdysozoans (Aguinaldo *et al.* 1997). Though not common today, priapulids were a comparatively abundant component of Cambrian marine faunas (see Conway Morris 1977a, 1998; Conway Morris & Robison 1986). The monophyly of the Priapulida has been unambiguously supported (Wills 1998b).

Genus *Palaeopriapulites* Hou, Bergström, Wang, Feng & Chen, 1999

Palaeopriapulites parvus Hou, Bergström, Wang, Feng & Chen, 1999

This is one of the rarer priapulid worms of the Chengjiang biota, known from a few delicately preserved compressed specimens.

Individuals are less than 10 mm long. The anterior proboscis and the posterior trunk are approximately equal in size and oval-shaped in lateral view. The proboscis bears an array of tiny spines arranged in about 20 longitudinal rows, and the trunk displays a number of compressed, longitudinal wrinkles. The trunk was possibly encased in a kind of thick cuticle, as is the case in some Recent priapulid larvae, where such cuticle is called the lorica. The gut is typically narrow and straight, but in one specimen it is curved near the posterior end of the trunk, a region surrounded by hooks.

The monotypic *Palaeopriapulites* was established within its own family, the Palaeopriapulitidae. *Priapulites* Schram, 1973, from the Carboniferous of Illinois, has a similar general morphology.

Like Recent and other Cambrian priapulids from the Burgess Shale, *P. parvus* is presumed to be a burrower. There is no direct evidence to indicate whether it was a predatorial carnivore or a digester of sediment.

The species is known only from the Lower Cambrian of Yunnan Province. Originally described from the Chengjiang area, it has recently also been reported from Anning, Yunnan Province (Zhang *et al.* 2001).

Key References Conway Morris 1977a, Hou *et al.* 1999.

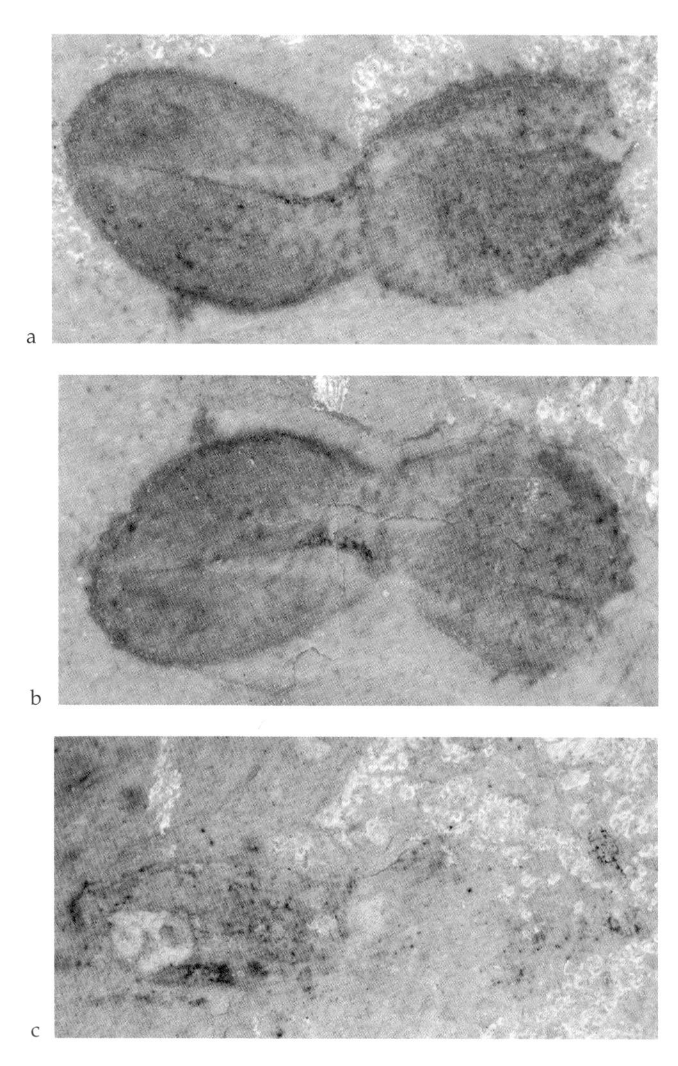

Figure 12.1 *Palaeopriapulites parvus*. (a) Lateral view, part (NIGPAS 115446a), ×7.0; Maotianshan. (b) Lateral view, counterpart (NIGPAS 115446b), ×7.0. (c) Lateral view (RCCBYU 10228), ×9.3; Maotianshan.

Genus *Protopriapulites* Hou, Bergström, Wang, Feng & Chen, 1999

Protopriapulites haikouensis Hou, Bergström, Wang, Feng & Chen, 1999

P. haikouensis is a relatively common species of worm, known from many tens of specimens. The material consists of delicate, soft-bodied compression fossils, including complete individuals, which are detected mainly by subtle color differences against the rock matrix.

The largest individuals are about 10 mm long. The anterior proboscis and the short, elongate oval-shaped posterior trunk are of subequal size. When preserved everted the narrower, anterior part (pharynx) of the proboscis is seen to bear a regular array of tiny spines, as does the broader base of the proboscis. The trunk has a series of longitudinal external ridges, of which 13–15 are visible in one specimen. The internal part of the trunk is occupied almost entirely by what is interpreted as a multi-coiled gut.

P. haikouensis is similar to the coeval *Palaeopriapulites parvus* from the Chengjiang area, from which it differs most obviously by having a coiled gut in the trunk region. Both species were established within monotypic genera and the new Family Palaeopriapulitidae. The small size of *P. haikouensis* and *P. parvus*, noticeable especially in the length of the trunk, contrasts with most other Recent and fossil priapulids. The Chengjiang biota species *Sicyophorus rarus*, from Haikou, originally described under uncertain taxonomic affinity, is probably a synonym of *P. haikouensis*.

It is likely that *P. haikouensis* was a burrower, but possibly not a predator. Some specimens have the gut filled with what appears to be mud, indicating that these individuals were perhaps digesting organic material from the sediment.

As alluded to by its name, this worm was originally collected from the Lower Cambrian near Haikou. It is unknown elsewhere.

Key Reference Hou *et al.* 1999.

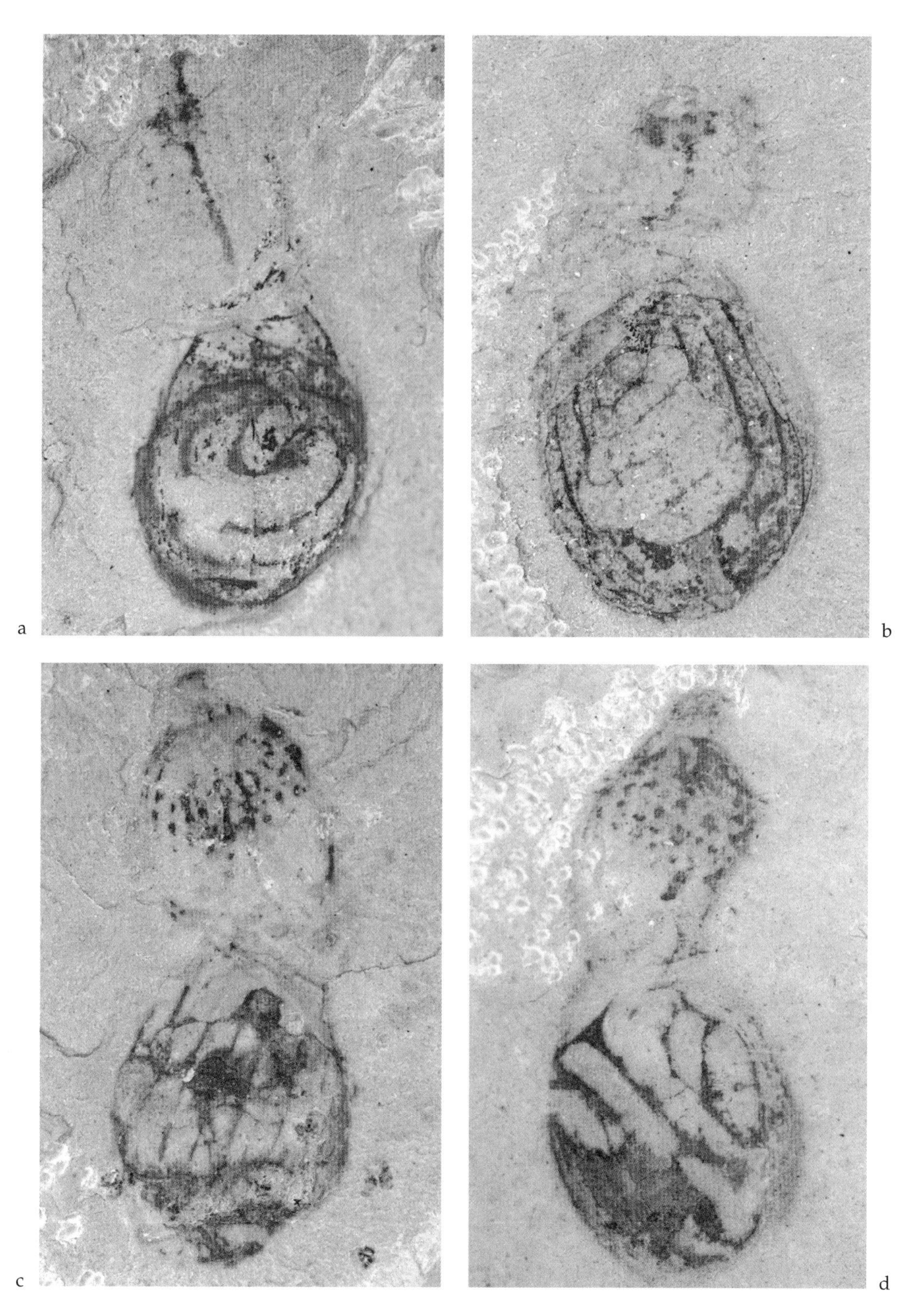

Figure 12.2 *Protopriapulites haikouensis*. (a) Lateral view (RCCBYU 10229), ×10.0; Mafang. (b) Lateral view (RCCBYU 10230), ×13.0; Mafang. (c) Lateral view (RCCBYU 10231), ×10.0; Mafang. (d) Lateral view (RCCBYU 10232), ×10.0; Mafang.

Genus *Acosmia* Chen & Zhou, 1997

Acosmia maotiania Chen & Zhou, 1997

This is a rare species. Specimens are preserved as flat impressions and many show a clear longitudinal gut trace.

Specimens are up to 45 mm long and 9 mm wide, and were probably cylindrical in life. The original description noted an anterior, barrel-shaped proboscis separated by a distinct constriction from an annulated trunk. The proboscis is armed with an array of hook-like projections. Papillae are evident in the anterior and posterior portions of the body, regularly spaced and symmetrical across the trunk; there is also an ornament of fine spines. The gut is straight, has a striated surface and occupies a third of the width of the trunk. The gut appears to be offset from the mid-line at the posterior end, perhaps indicating a posteroventral opening.

Based on the presence and morphological details of the proboscis and trunk Chen & Zhou (1997) considered that *Acosmia* is a priapulid worm, the largest in the Chengjiang fauna. Though more circumspect about its affinity, Hou *et al.* (1999) assigned it to a new family, the Acosmiidae.

Some specimens of *A. maotiania* are preserved tightly curved or with the anterior part erect, suggesting that it might have lived in a U-shaped burrow. Some individuals have a broader alimentary canal, which the original authors interpreted as food filling, and this may indicate that *A. maotiania* was a deposit feeder.

The monospecific *Acosmia* is unknown outside the Chengjiang Lagerstätte.

Key References Chen & Zhou 1997, Hou *et al.* 1999.

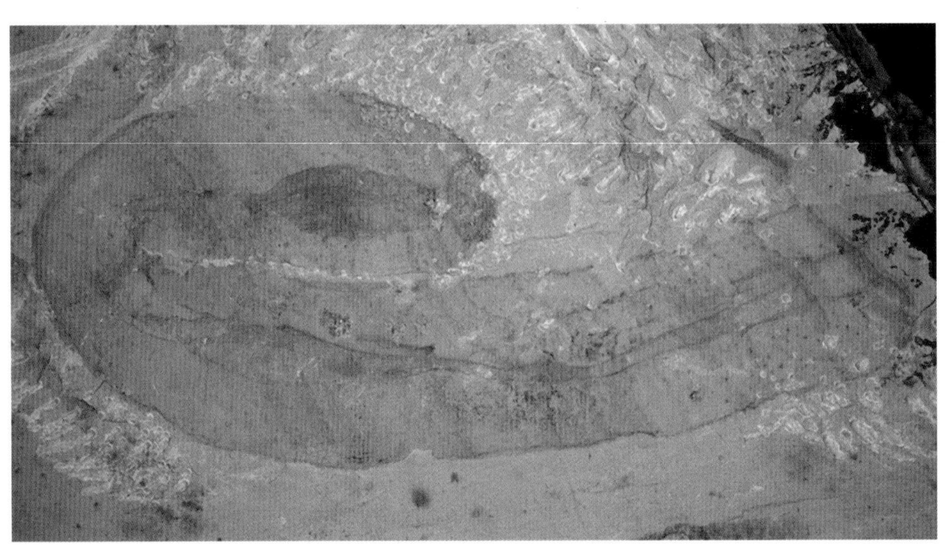

a

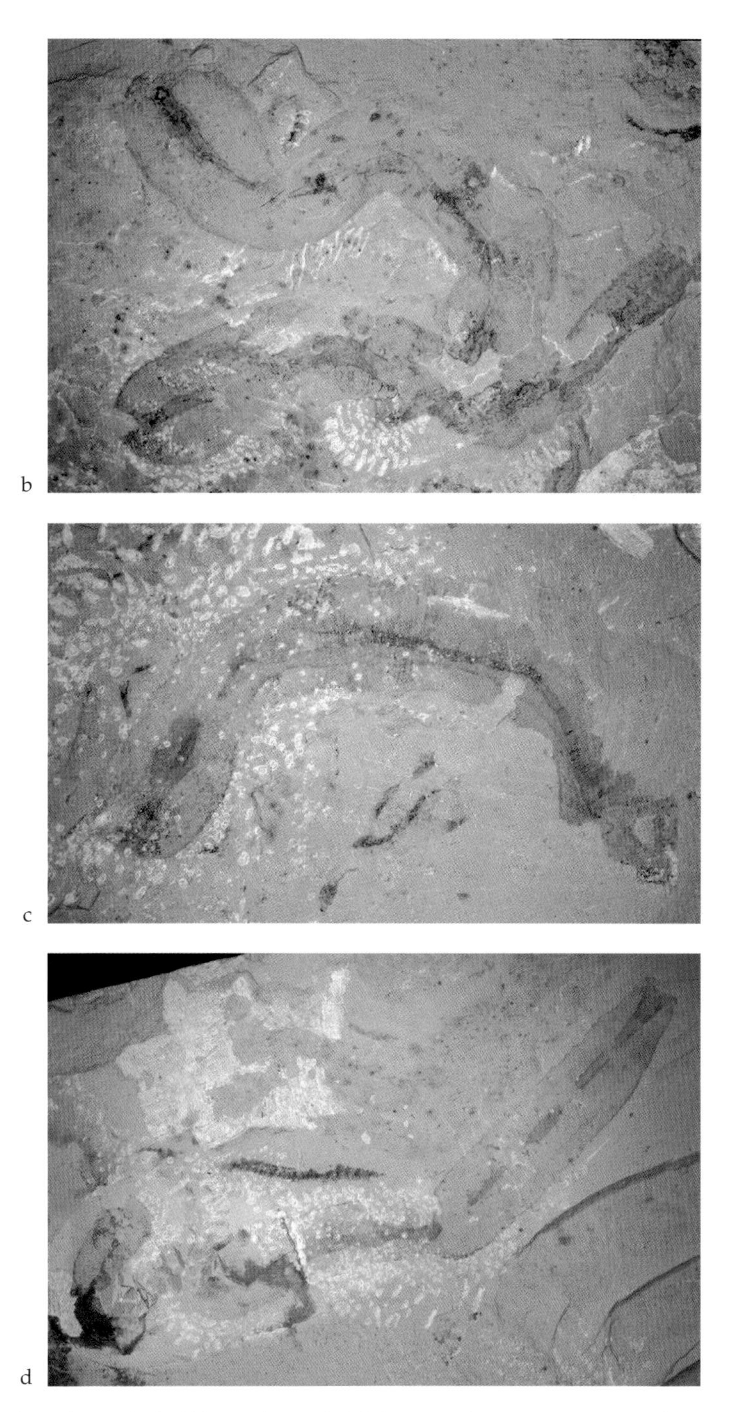

Figure 12.3 *Acosmia maotiania*. (a) Lateral view (RCCBYU 10233), ×3.1; Maotianshan. (b) Lateral view (RCCBYU 10234), ×3.5; Maotianshan. (c) Lateral view (RCCBYU 10235), ×3.3; Maotianshan. (d) Lateral view (RCCBYU 10236), ×2.1; Maotianshan.

Genus *Paraselkirkia* Hou, Bergström, Wang, Feng & Chen, 1999

Paraselkirkia jinningensis Hou, Bergström, Wang, Feng & Chen, 1999

This priapulid worm is known from hundreds of specimens. The fossils typically consist of an outer tube from which protrudes a spiny proboscis. Both the soft parts and the more decay-resistant organic tube are preserved as compression fossils.

In some specimens the proboscis is fully everted and is clearly divided transversely into several parts, each characterized by spines of a particular size and array. The narrow, elongate tapering tube, which presumably housed the trunk of the animal, has a total length of about 20 mm and a width of about 3 mm. The surface of the tube displays fine, regularly spaced annulations, representing incremental growth.

The monotypic *Paraselkirkia* is a member of the Selkirkiidae, a family established on Burgess Shale material. The Chengjiang worm *Selkirkia sinica* bears close resemblance to *P. jinningensis* and may be the same species.

It is reasonable to assume that *P. jinningensis* had a lifestyle like the morphologically similar *Selkirkia columbia* Conway Morris, 1977 from the Burgess Shale. The latter species, in which the tube is open at both ends, is interpreted as a burrowing, tube-dwelling carnivore. It may have been orientated vertically in the sediment or arranged head-down at various angles with the posterior-most tip of the tube projecting into the water to facilitate waste disposal (see Briggs *et al.* 1994, Conway Morris 1998).

P. jinningensis is known only from the Chengjiang biota.

Key References Conway Morris 1977a, Conway Morris 1998, Hou *et al.* 1999.

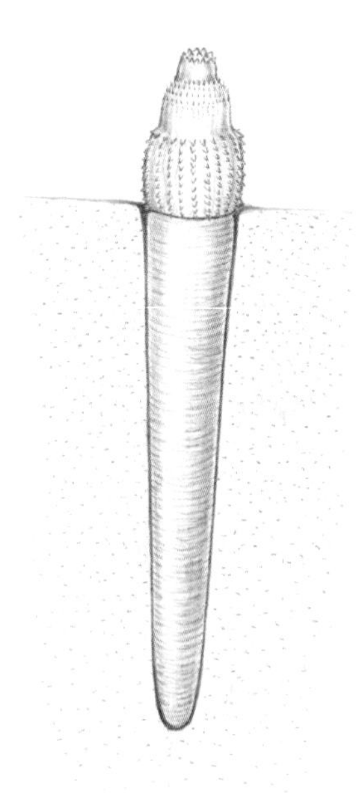

Figure 12.4 Reconstruction of *Paraselkirkia jinningensis* (based in part on the reconstruction of the Burgess Shale *Selkirkia* in Briggs *et al.* 1994).

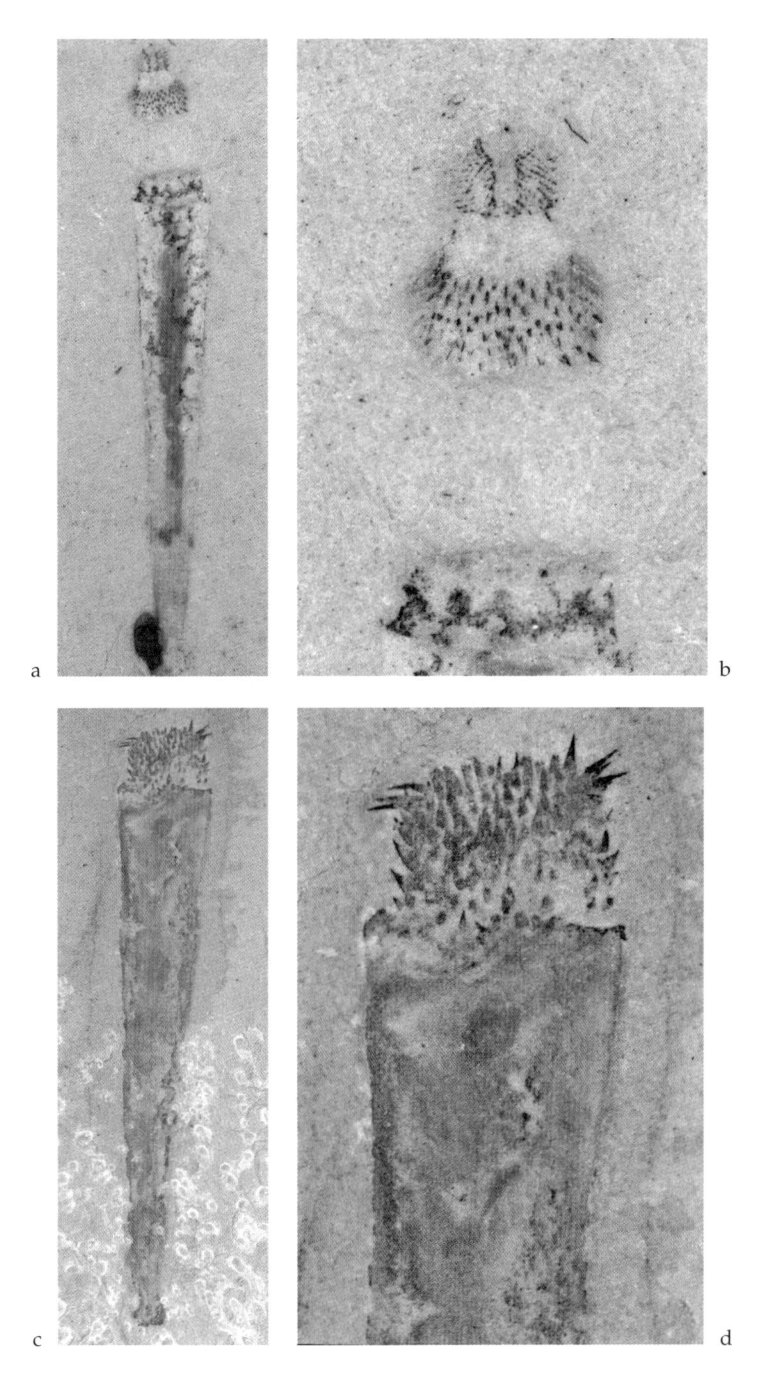

Figure 12.5 *Paraselkirkia jinningensis*. (a) Lateral view (RCCBYU 10237), × 6.2; Ercaicun. (b) Detail of everted proboscis (RCCBYU 10237), × 18.6. (c) Lateral view (RCCBYU 10238), × 7.6; Ercaicun. (d) Detail of partly everted proboscis (RCCBYU 10238), × 19.0.

Genus *Archotuba* Hou, Bergström, Wang, Feng & Chen, 1999

Archotuba conoidalis Hou, Bergström, Wang, Feng & Chen, 1999

This reasonably common species is known essentially from its tubes, which are preserved as two-dimensional, whitish or reddish-brown films picked out against the typically buff-colored rock matrix.

The tube is elongate cone-shaped, with a maximum length of just over 50 mm and a maximum diameter of over 6 mm. Most specimens lack any sign of ornament, but a few specimens show surface annulations, presumably representing marginal incremental growth. A dark axial structure along the middle part of some tubes may indicate the trace of the intestine (Hou *et al.* 1999).

Though no diagnostic soft parts of this animal were found, the original authors assumed that the tubes represent part of a tube-dwelling priapulid worm in the manner of Walcott's (1911c) Burgess Shale *Selkirkia* and the Chengjiang genus *Paraselkirkia*. If *A. conoidalis* is a priapulid, its mode of life is problematic. There are examples in which several of these tubes have been found together, orientated subparallel with their apices emanating from near one another and in some cases touching, as if fixed, to a brachiopod, hyolithid or other type of shell. Most fossil priapulids are considered to be active burrowers and sedentary lifestyles are invoked for only a few species. In contrast, others have interpreted this species as a sessile benthic cnidarian (*Cambrorhytium* sp. nov. of Chen & Zhou 1997; see also Chen *et al.* 1996), although there is no fossil evidence for the presence of the tentacles depicted in those alternative reconstructions.

The Chengjiang taxon *Selkirkia*? *elongata* is an elongate, cone-shaped fossil closely similar to *A. conoidalis*, and these species may be synonymous.

All known occurrences of *A. conoidalis* are from the Lower Cambrian of Yunnan Province.

Key References Chen *et al.* 1996, Chen & Zhou 1997, Hou *et al.* 1999.

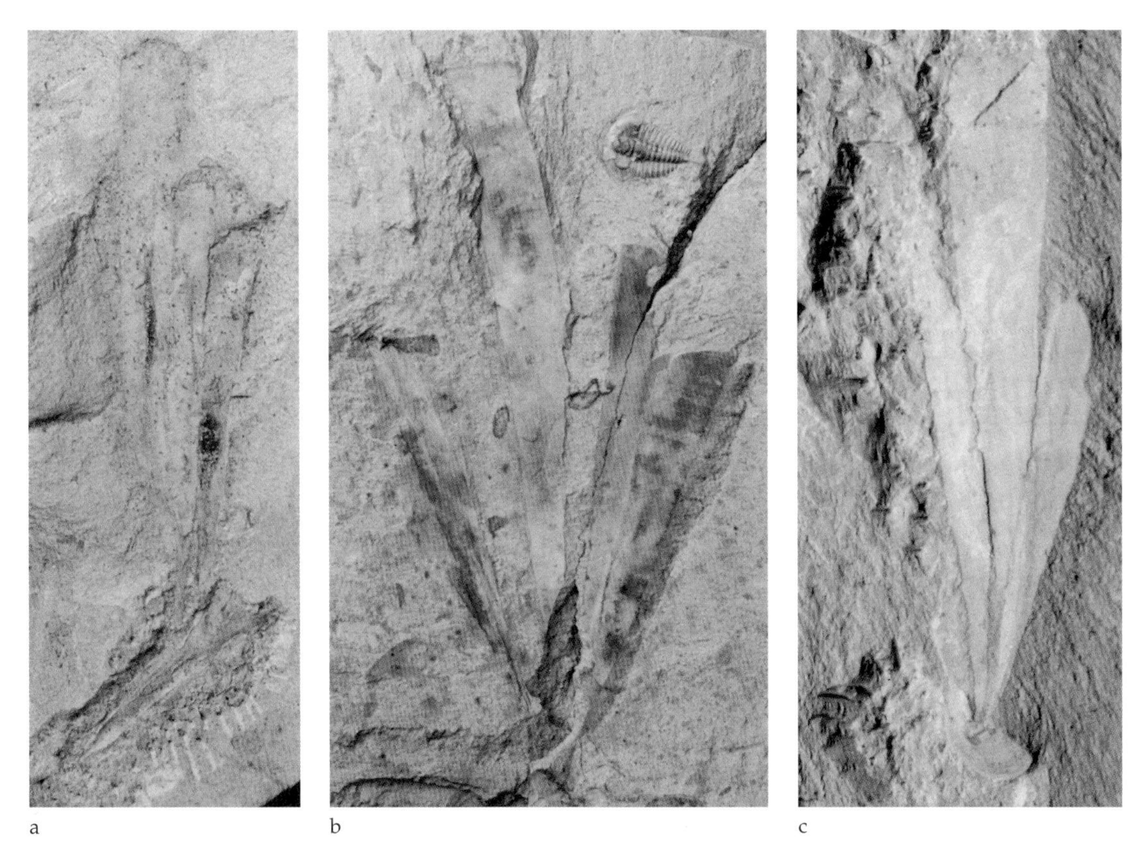

Figure 12.6 *Archotuba conoidalis*. (a) Lateral view of three specimens, seemingly attached to a hyolithid (NIGPAS 115452), ×1.6; Ma'anshan. (b) Lateral view of several specimens, seemingly attached to a hyolithid fragment (the trilobite, upper right, is *Eoredlichia*) (NIGPAS 115450a), ×2.3; Ma'anshan. (c) Lateral view of three specimens, seemingly attached to a brachiopod (NIGPAS 115453), ×4.7; Maotianshan.

13 PHYLUM HYOLITHA

Hyolitha occur exclusively as fossils, from the Cambrian to Permian, and are of problematic affinity. Some authors regard them as a separate phylum. They had elongate, cone-shaped, calcareous shells (see Feng *et al.* 2001), but little is known about their soft parts. Two main groups have been recognized, namely the hyolithids and the orthothecids (Marek 1966). Hyolithids are protostome bilaterians, and may be related to sipunculans (the unsegmented, marine "peanut worms") and the molluscs, the group that includes bivalves, gastropods and cephalopods (Martí Mus & Bergström 2002).

From what is known, for example, from Middle Cambrian material from Utah and the Burgess Shale (see Briggs *et al.* 1994, Conway Morris 1998), in life the shell of hyolithids had an aperture covered by a lid-like operculum and a pair of narrow curved struts (helens) projecting between the conch and operculum. Both the conch and operculum bear a variety of muscle scars. Hyolithids are generally considered to have lived on the substrate, resting on the flatter side of the shell for stability. The helens might also have aided stability in soft sediment and could perhaps have been used to facilitate what was possibly a slow, labored form of locomotion. Many species appear to have favored living on mud substrates.

Cambrian hyoliths are known, for example, from Siberia, North America and many European countries (see Malinky & Berg-Madsen 1999). Hyoliths are fairly common fossils of the Chengjiang biota; quite a number of species have been described or simply recorded, several of which are difficult to evaluate based on the published information.

Genus *Linevitus* Syssoiev, 1958

Linevitus opimus Yu, 1974

This relatively rare hyolithid species is known from the remains of its thin, mineralized shells, which typically are preserved flattened but retaining slight relief.

The elongate, cone-shaped conch is up to 15 mm wide at its apertural end and up to 30 mm long, tapering gradually to a sharp apex. The apical angle of the shell is approximately 33°. The external surface of the shell shows fine growth lines, parallel to the margin of the aperture. The dorsal side is slightly convex and has a central longitudinal groove that tapers from aperture to apex.

L. opimus is known from about ten specimens as associates of the soft-bodied elements of the Chengjiang biota. It may be more common at other horizons locally.

Key References Yu 1974, Hou *et al.* 1999.

Figure 13.1 *Linevitus opimus*. Dorsal view of shell (RCCBYU 10239), ×5.4; Maotianshan.

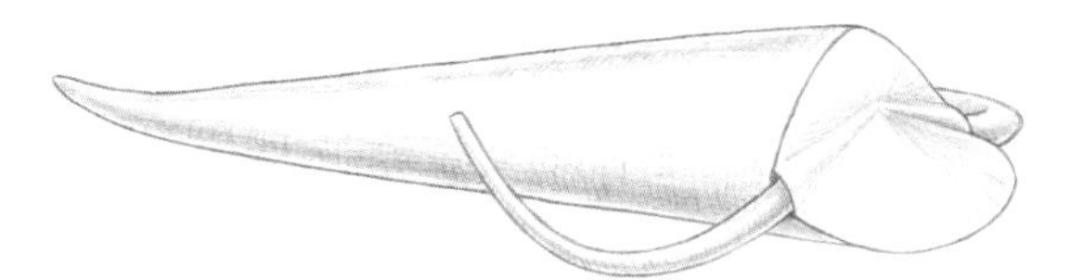

Figure 13.2 Generalized reconstruction of a hyolithid.

Genus *Burithes* Missarzhevsky, 1969

Burithes yunnanensis Hou, Bergström, Wang, Feng & Chen, 1999

B. yunnanensis is known from many tens of shells. They are normally preserved flattened yet retaining an overall cone shape and, in some cases, weak relief. Rare specimens bearing partial remains of the soft tissues await description.

Of all the Chengjiang hyolithids, this species has the largest shell. It attains a length of up to 35 mm and is up to 15 mm wide at the apertural end. The apical angle of the shell is about 25°. The shell surface has weak growth lines parallel to the apertural margin, but lacks any form of longitudinal ornament.

The Chengjiang hyolithid *Glossolites magnus* is closely similar to *B. yunnanensis* and may represent the same species.

All known occurrences of *B. yunnanensis* are from the Lower Cambrian of Yunnan Province.

Key Reference Hou *et al.* 1999.

Figure 13.3 *Burithes yunnanensis*. Flattened shell (NIGPAS 115438), ×4.0; Maotianshan.

Genus *Ambrolinevitus* Sysoiev, 1958

Ambrolinevitus maximus Jiang, 1982

The shells of this hyolithid are rare as associates of the soft-bodied biota but are more abundant at other horizons. They are preserved as flattened, essentially two-dimensional specimens, with little relief.

The shell is elongate, cone-shaped and has straight sides. It has a maximum length of 16 mm and its maximum width at the aperture is 6 mm; the apex is somewhat obtuse. Many shells show a narrow longitudinal central groove that probably represents a post-mortem "collapse" feature. Narrow, dense growth lines, some 19–22 per millimeter, parallel the apertural margin.

Some closely packed and in some cases subparallel aligned bedding plane assemblages of *Ambrolinevitus* material in the Chengjiang biota have been interpreted as possible gut or coprolite contents (Chen & Zhou 1997).

A. maximus is unknown outside the Chengjiang biota.

Key References Jiang 1982, Chen *et al.* 1996, Chen & Zhou 1997, Hou *et al.* 1999.

Figure 13.4 *Ambrolinevitus maximus*. Flattened shell (RCCBYU 10240), ×8.4; Xiaolantian.

Ambrolinevitus ventricosus Qian, 1978

This small hyolithid species occurs in abundance in the Chengjiang biota. It is not uncommon to find tens of specimens crowded together, in some cases showing a sub-parallel, possible paleocurrent, orientation. The moldic shells, originally composed of calcium carbonate, are preserved flattened with some slight relief. Soft parts are unknown.

The cone-shaped shell has straight sides, a sharply pointed apex and is relatively small. The largest specimens are only about 5 mm long and 1.5 mm wide at the apertural end. The angle of divergence of the shell is approximately 20°. The external part of the shell has narrow dense growth lines, some 18–20 per millimeter. A central longitudinal groove, representing a probable post-mortem "collapse" feature, is common.

A. ventricosus has been recorded only from the Lower Cambrian of Yunnan Province.

Key References Qian 1978, Hou *et al.* 1999.

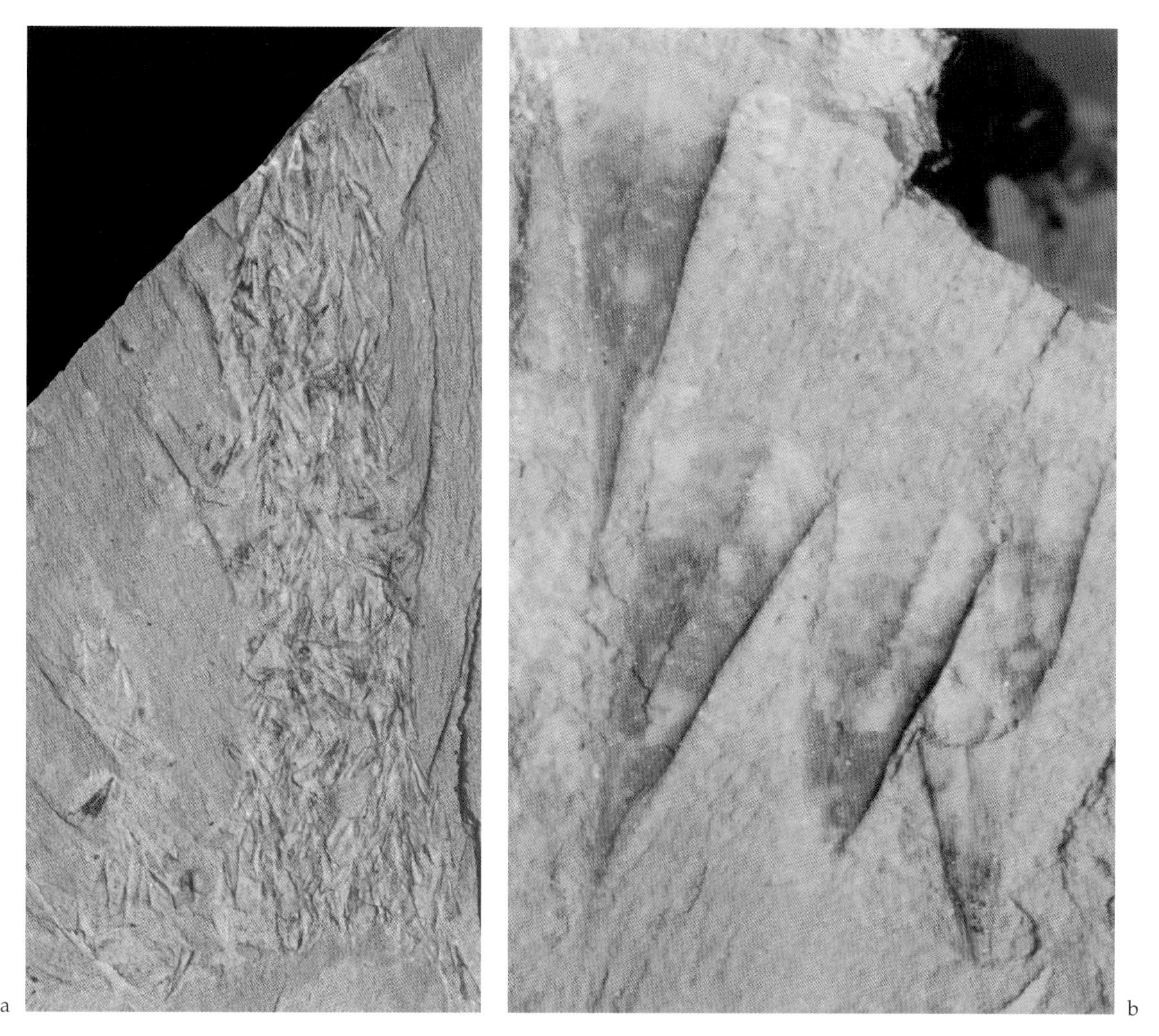

Figure 13.5 *Ambrolinevitus ventricosus*. (a) Slab containing more than one hundred specimens (RCCBYU 10241), ×1.9; Ma'anshan. (b) Detail of the slab, showing five aligned specimens (RCCBYU 10241), ×8.9.

14 PHYLUM LOBOPODIA

Lobopodians are small marine and terrestrial animals that have been colloquially termed "velvet worms" or "worms with legs". Their vermiform body, consisting of soft tissue and lightly sclerotized cuticle, bears uniramous appendages. Some fossil forms have isolated, partially mineralized plates in the trunk region. The Recent species, members of the Onychophora, show morphological differences from most fossil representatives—for example the possession of antennae and a ventral mouth with jaws. Lobopodians have been allied in particular with arthropods, figuring prominently in assessments of the origins and relationships of this major invertebrate group.

Most fossil lobopodians were marine and they are known almost entirely from Cambrian rocks, the only others being from Carboniferous and (Tait 2001) Tertiary deposits. Cambrian lobopodians, which occur in several parts of the world from North America to China, were morphologically more varied and formed a relatively more significant component of the fauna than do their present-day counterparts. All Recent species are terrestrial; the morphology of the Carboniferous *Helenodora* Thompson & Jones (1980) from the USA suggests that the move on to land had been accomplished by then.

Some 20 named and unnamed lobopodian species are known from the fossil record. Six named genera, each with a single species, come from the Cambrian of Yunnan Province, making it the richest source of fossil lobopodians. Another new but as yet undescribed form has recently been reported from the province (Zhang *et al.* 2001). The nature of *Megadictyon haikouensis* and *Tylotites petiolaris*, supposed lobopodians from the Cambrian of the Chengjiang area (Luo & Hu *in* Luo *et al.* 1999), is uncertain and they are not included here.

Genus *Luolishania* Hou & Chen, 1989

Luolishania longicruris Hou & Chen, 1989

This is a rare lobopodian of rather generalized appearance. It was originally described on the basis of a matching part and counterpart; subsequently five further specimens have been recovered. Specimens are preserved as essentially flattened impressions, though with some slight relief.

The length of complete specimens is estimated to be about 15 mm. The head is rounded and slightly elongate, separated from the trunk by a slight constriction. The trunk is relatively long and slender; posteriorly it is narrower and tapers to a short, bluntly rounded projection. There are 16 pairs of annulated legs, on some of which four or five claws have been identified. Each leg originates from a smooth band around the trunk, and dorsolaterally on each of these bands there is a pair of small, rounded bumps that are interpreted as sclerites. A raised structure on the dorsal mid-line between some pairs may be a third sclerite. Between the bands 6–8 annuli have been described in the mid-length of the trunk, decreasing to three in both the anterior and posterior regions, though these numbers are uncertain because of the possibility of displaced, overlapping annuli (Ramsköld 1992). Many of the annuli carry small nodes. Traces of the gut are evident.

L. longicruris is the only known species of *Luolishania*. This genus has been considered to be closely related to *Xenusion* Pompeckj, 1927, an Early Cambrian lobopodian from a glacial erratic in Germany, and also to *Aysheaia* Walcott, 1911 from the Burgess Shale (Hou & Bergström 1995, Ramsköld & Chen 1998). One analysis places it in a group including these other two genera together with *Onychodictyon* (Ramsköld & Chen 1998), which is found in Yunnan Province.

L. longicruris is found with sponges. This association, together with its possession of clawed legs, has led to the suggestion that it crawled up and preyed on them—a lifestyle like that envisaged for *Aysheaia*. The species is known only from the Lower Cambrian of Yunnan Province.

Key References Hou & Chen 1989b, Ramsköld 1992, Hou & Bergström 1995, Ramsköld & Chen 1998, Bergström & Hou 2001.

Figure 14.1 *Luolishania longicruris*. Dorsolateral view (RCCBYU 10242), ×9.8; Mafang.

Genus *Paucipodia* Chen, Zhou & Ramsköld, 1995

Paucipodia inermis Chen, Zhou & Ramsköld, 1995

A comparatively featureless lobopodian, this species was established on the basis of four specimens. Recently it has been discussed briefly and reconstructed anew from other specimens that overall are similar to those first described, but which are apparently more complete, are at least twice as long with a long posterior termination, and which show three more leg pairs (Bergström & Hou 2001). Photographs of the latter type of specimen are, for the first time, included here; the material will be fully described elsewhere (Hou *et al.* in press). The first-described specimens have been compacted to a single extremely thin layer, with very low relief present in the claws and trunk annuli.

Specimens can reach at least 80 mm in length. The head tapers anteriorly to a rounded margin, and the mouth is indicated in a terminal or near-terminal position. Fine annuli are present on the trunk, which otherwise is rather featureless, lacking any papillae or paired sclerites. An approximately centrally positioned gut runs along the whole length of the body, though its position shows variability that probably reflects the post-mortem decay of supporting soft tissue. The trunk is considerably extended behind the last leg pair, of which nine have now been recognized. Circular structures positioned low on the side of the trunk in the original specimens have been interpreted as leg attachment sites. Each of the legs has two broadly based, sharp claws.

Just one species of *Paucipodia* has been described. The genus has been placed in a lobopodian group that also includes *Hallucigenia*, *Microdictyon* and *Cardiodictyon*, on the basis of head, leg, and annulation characters (Ramsköld & Chen 1998).

This soft-bodied marine animal with clawed legs has, in common with other Cambrian lobopodians, been interpreted as epibenthic in habit, crawling on the seafloor and on other organisms. In particular *Paucipodia* is found on or near specimens of *Eldonia*, this occurrence probably representing a life association. Its claws would certainly aid attachment, and they were perhaps also used on prey or carrion.

The Chengjiang area yielded the initial finds of *P. inermis*, with the Haikou district providing some of the later material. *Paucipodia* is known with certainty only from the Lower Cambrian of Yunnan Province; it has questionably been recorded from the Middle Cambrian Kaili Lagerstätte of Guizhou Province (Zhao *et al.* 1999a).

Key References Chen *et al.* 1995c, Bergström & Hou 2001, Hou *et al.* in press.

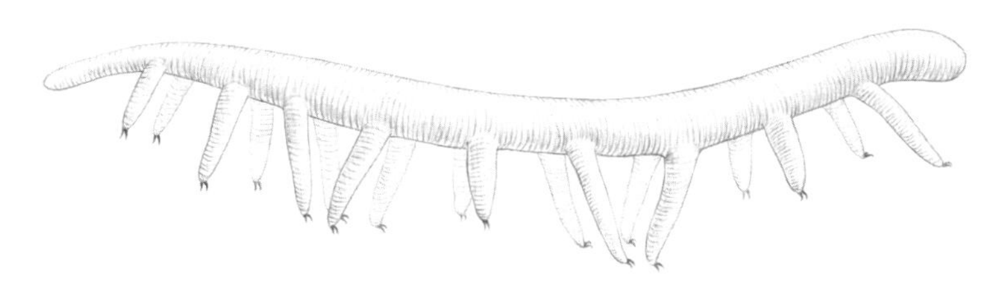

Figure 14.2 Reconstruction of *Paucipodia inermis* (after Bergström & Hou 2001).

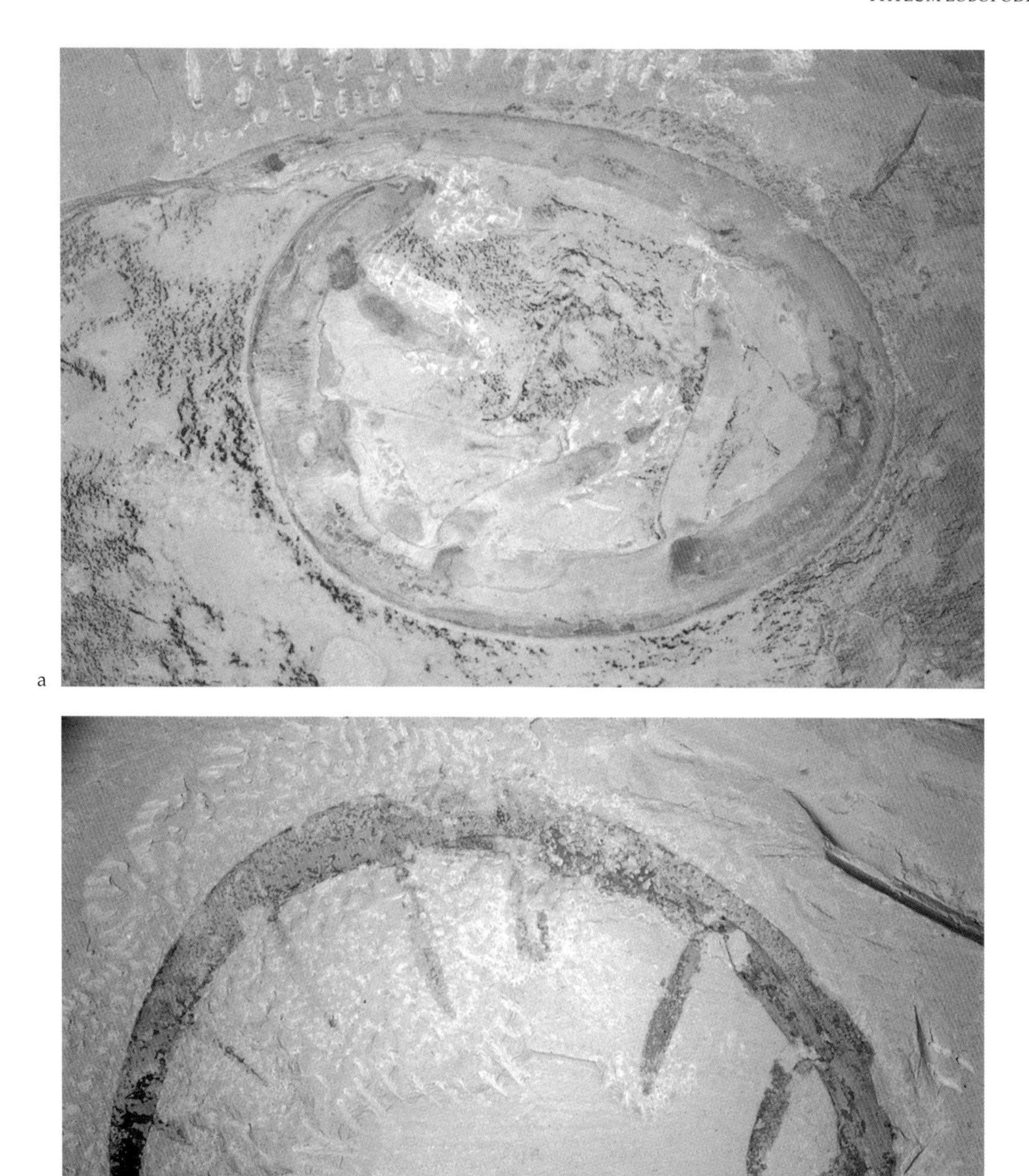

Figure 14.3 *Paucipodia inermis*. (a) Lateral view (RCCBYU 10185), ×3.2; Mafang. (b) Lateral view (RCCBYU 10183), ×2.4; Mafang.

Genus *Cardiodictyon* Hou, Ramsköld & Bergström, 1991

Cardiodictyon catenulum Hou, Ramsköld & Bergström, 1991

This is one of the more commonly represented lobopodians in the Chengjiang fauna, the number of specimens reaching at least double figures. Specimens are flattened, and sometimes show very slight relief, including a three-dimensionally preserved gut.

Complete, large specimens are about 20 mm long. The head is elongate, probably covered with a sclerite or pair of sclerites, and with a rounded termination. The trunk is long, slender, tapering slightly anteriorly and posteriorly and annulated. Complete specimens have 23 to 25 angular, paired sclerites with raised marginal ridges regularly placed along the trunk. Sclerite morphology varies: in smaller specimens they are ventrally pointing, narrow V-shaped structures, and in larger adults they are more shield-like. The paired sclerites appear to meet dorsally, though recently it has been noted that they may also be interpreted as extending dorsally over the mid-line (Bergström & Hou 2001). Ventrally, beneath each of the successive sclerite pairs, there is a pair of legs. Each leg has numerous, fine annuli and curved, pointed claws—two according to one study (Ramsköld & Chen 1998), though four or possibly five according to another (Hou & Bergström 1995). The same studies also admit to a slightly different number of appendage pairs in the head region in front of the most anterior pair of sclerites, the former study recognizing perhaps one or two, the latter two or three. The posterior trunk termination behind the last plate pair is blunt. The gut trace is dark colored and extends along the length of the animal.

C. catenulum represents the only known species of *Cardiodictyon*, which is generally considered to belong to a group of lobopodians that includes *Hallucigenia* and *Microdictyon*. These three genera have been determined to have certain characters of the head, annuli, leg, and sclerites in common (Hou & Bergström 1995, Ramsköld & Chen 1998). *Paucipodia* is also included in this group, though unlike the others it lacks paired sclerites.

As with the other lobopodians, *Cardiodictyon* is interpreted as part of the epibenthos. It is unknown outside Yunnan Province.

Key References Hou *et al.* 1991, Ramsköld 1992, Hou & Bergström 1995, Ramsköld & Chen 1998, Bergström & Hou 2001.

Figure 14.4 Reconstruction of *Cardiodictyon catenulum* (modified from Bergström & Hou 2001).

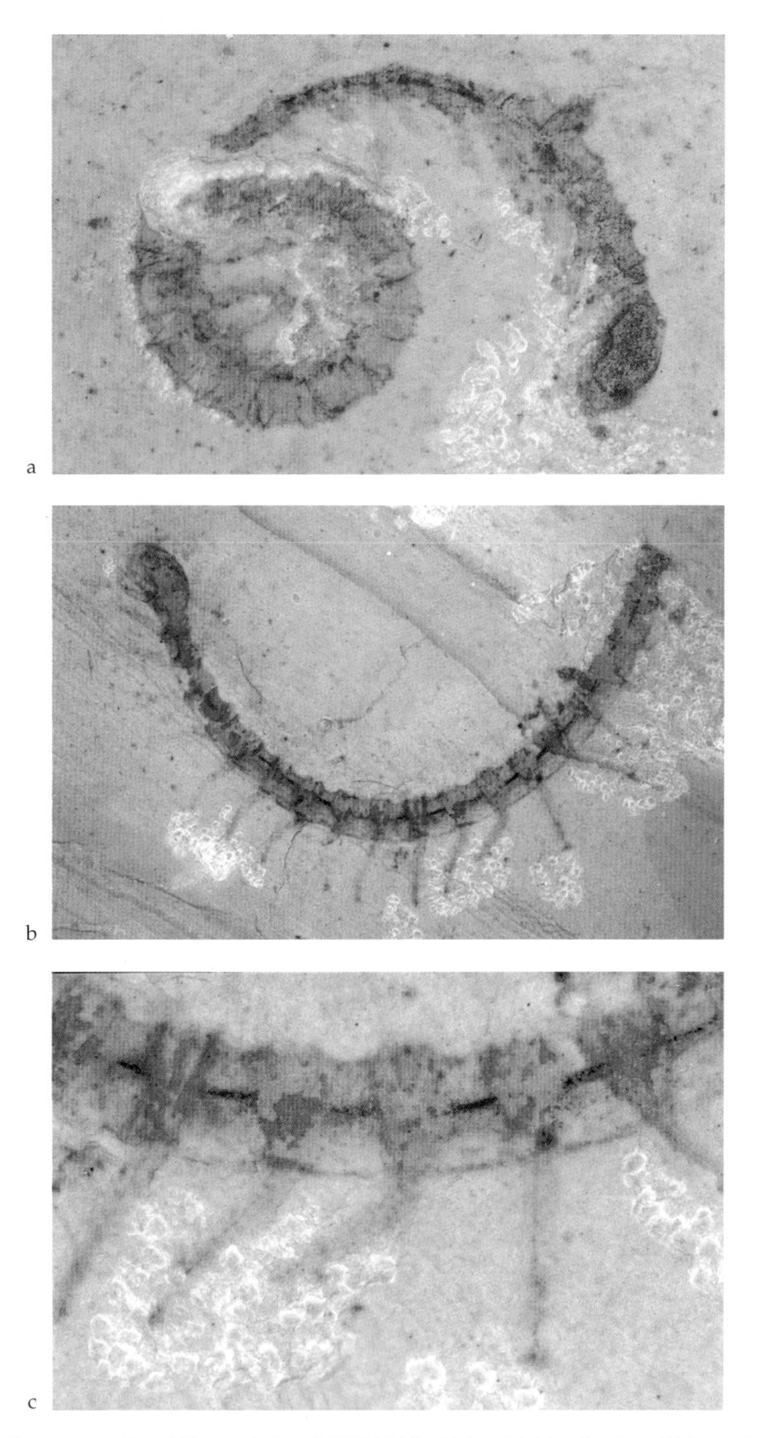

Figure 14.5 *Cardiodictyon catenulum*. (a) Lateral view (NIGPAS 115293a), ×7.3; Maotianshan. (b) Lateral view (RCCBYU 10245), ×4.5; Mafang. (c) Detail of trunk region (RCCBYU 10245), ×13.5.

Genus *Hallucigenia* Conway Morris, 1977

Hallucigenia fortis Hou & Bergström, 1995

The original description of *H. fortis* was based on two specimens. A further 15 have been reported subsequently, but not fully described or interpreted.

One fairly complete specimen of this species is about 22 mm long. The head is ellipsoidal in shape and separated from the trunk by a constriction. According to one study, the head is covered by a pair of sclerites (Hou & Bergström 1995), though another investigation did not support this interpretation (Ramsköld & Chen 1998). Immediately behind the head there are two pairs of long, slender, annulated appendages that are positioned close together. The trunk has annuli, seven pairs of dorsolaterally sited sclerites and seven pairs of annulated legs. A pair of curved claws may be present on each leg. Each sclerite has a base that supports a moderately long spine. Posterior to the seventh sclerite and leg pair, there is an eighth leg pair without a corresponding sclerite pair. The trunk tapers posteriorly. Short sections of the gut are preserved in some specimens.

Hallucigenia contains one other known species, *H. sparsa* from the Middle Cambrian Burgess Shale, which was originally considered by Walcott (1911c) to belong to the polychaete worm genus *Canadia*. Revision of *Canadia sparsa* saw it placed as the type species of *Hallucigenia*, a genus initially interpreted as of enigmatic affinity before its true identity as a lobopodian became recognized. One analysis has *Hallucigenia* most closely related to *Microdictyon* on the basis of leg and anterior appendage characters, and to be combined with *Cardiodictyon* and *Paucipodia* in a larger grouping (Ramsköld & Chen 1998).

An epibenthic, crawling mode of life is envisaged for *Hallucigenia*. The Lower Cambrian of Yunnan Province is the only known provenance of *H. fortis*. However the genus, in addition to its occurrence in British Columbia and Yunnan Province, may also be present in the Kaili Lagerstätte of Mid-Cambrian age from Guizhou Province (Zhao *et al.* 1999a).

Key References Conway Morris 1977b, Hou & Bergström 1995, Ramsköld & Chen 1998, Bergström & Hou 2001.

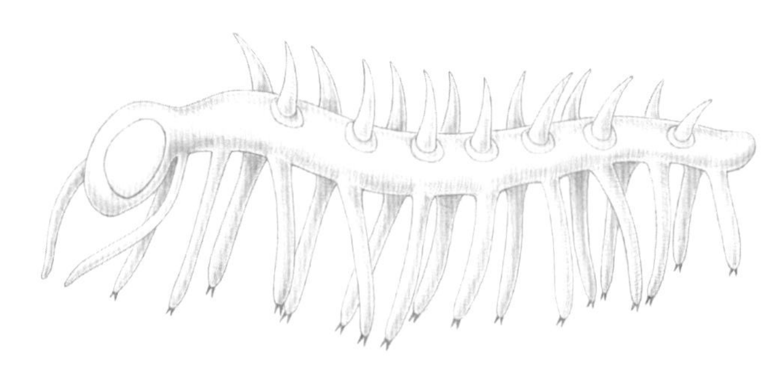

Figure 14.6 Reconstruction of *Hallucigenia fortis* (modified from Bergström & Hou 2001).

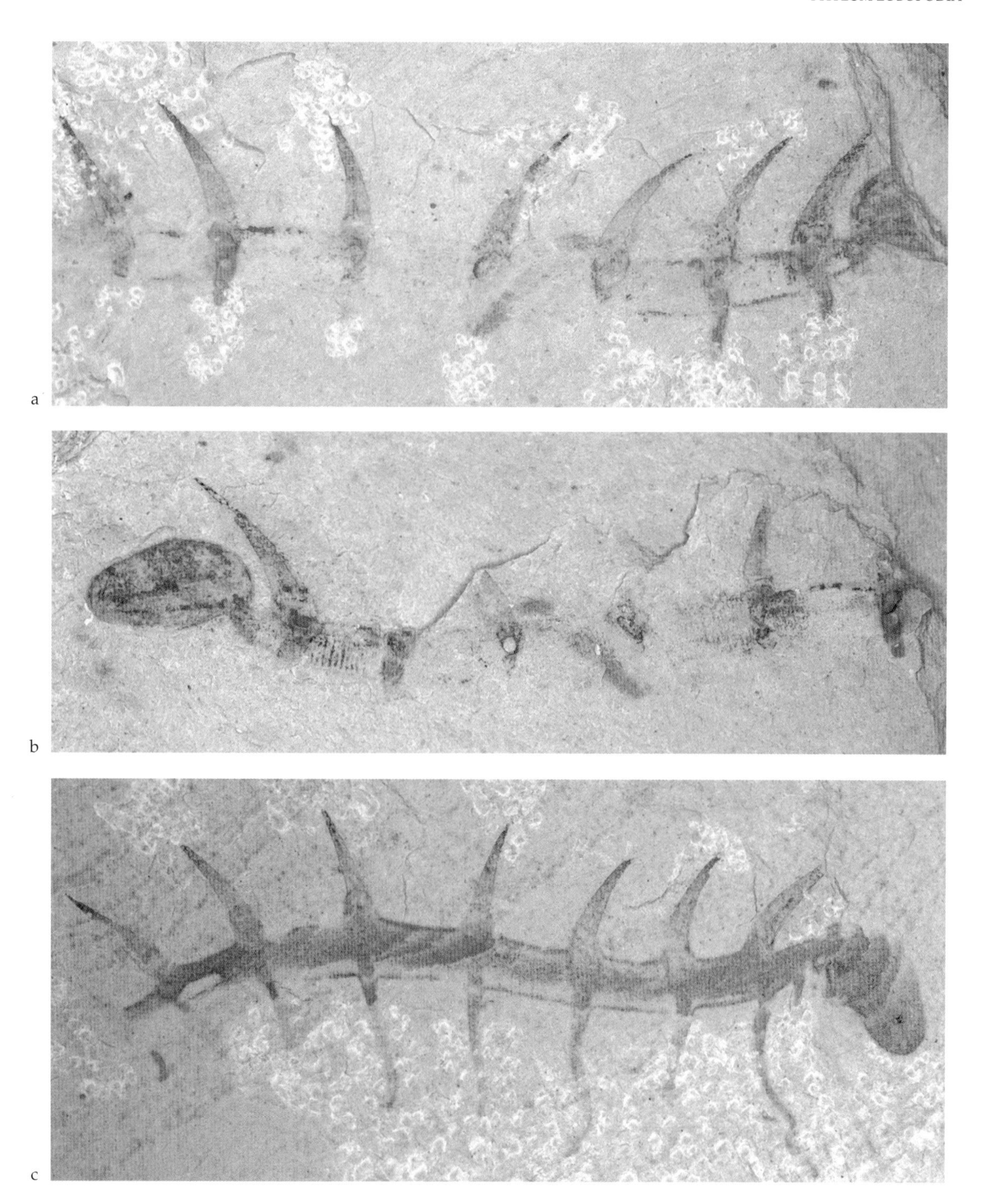

Figure 14.7 *Hallucigenia fortis*. (a) Lateral view (RCCBYU 10246), ×8.0; Mafang. (b) Lateral view (RCCBYU 10247), ×9.2; Mafang. (c) Lateral view (RCCBYU 10248), 8.4; Mafang.

Genus *Microdictyon* Bengtson, Matthews & Missarzhevsky, 1981

Microdictyon sinicum Chen, Hou & Lu, 1989

M. sinicum is known from at least 80 specimens. Many retain their gut—an indication that they are not molts.

Specimens range from less than 10 mm to about 77 mm long. The trunk is cylindrical. Interpretations differ as to which is the anterior, head end, and which is the posterior end. One end is long and narrow and tapers to a rounded termination, the other is very short and terminates in a small projection. There are nine pairs of trunk sclerites. Their size and shape are variable on a single specimen: relatively small and dorsoventrally compressed, round, ovoid or more rhomboidal. The surface of each plate has hexagonal, cylindrical perforations and spiky nodes. The trunk areas that bear the plate pairs are smooth and dorsally swollen, and between them the body is annulated. There are ten pairs of annulated appendages. Each pair lies directly beneath a sclerite pair except for the two appendage pairs at the short end of the body, which originate just anteriorly and posteriorly of the terminal sclerite pair. All appendages have a central canal and two distal claws.

Microdictyon was first established on the basis of isolated, phosphatic net-like plates of the then enigmatic *M. effusum* Bengtson *et al.* (1981) from the Lower Cambrian of Siberia. The later discovery of *Microdictyon sinicum* showed that similar plates formed part of a lobopodian. *Microdictyon* is closely related to *Hallucigenia* and *Cardiodictyon* and, more distantly, *Paucipodia* (Hou & Bergström 1995, Ramsköld & Chen 1998). *Quadratapora* Hao & Shu (1987) from the Lower Cambrian of Shaanxi Province and Siberia, known only from isolated sieve-like plates, is also apparently related to *Microdictyon*.

M. sinicum probably used its claws for attachment, onto *Eldonia* in particular (Chen *et al.* 1995d, Chen & Zhou 1997, Bergström & Hou 2001). *Microdictyon* perhaps fed on prey or carcasses (Bergström & Hou 2001), or may be it was a microphagous, nutrient particle feeder (Chen *et al.* 1995d). Also, its net-like sclerites have been likened to compound eyes and, possibly, to be homologous with arthropod eyes (Dzik 2003).

Microdictyon includes at least ten species (some unnamed) from the Lower and Middle Cambrian of North and Central America, Europe, Asia and Australia. All records outside China are based on sclerites. It occurs in China in the Lower Cambrian of Yunnan and (Tong 1989, Li & Zhu 2001) Shaanxi provinces, and apparently in the Middle Cambrian of Guizhou Province (Zhao *et al.* 1999a). *M. sinicum* is known only from the Chengjiang area.

Key References Chen *et al.* 1989a, Chen *et al.* 1995d, Hou & Bergström 1995, Bergström & Hou 2001.

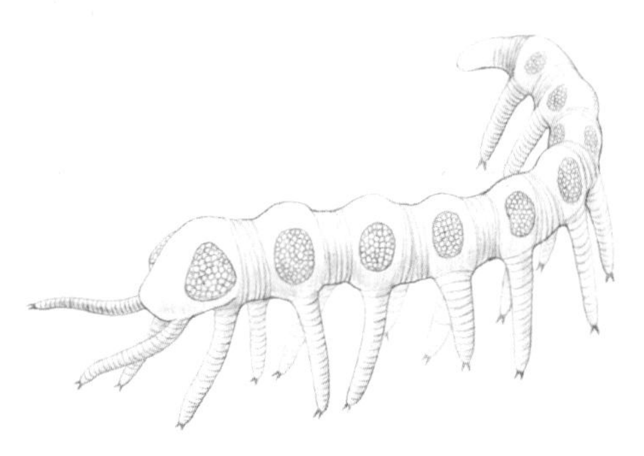

Figure 14.8 Reconstruction of *Microdictyon sinicum* (after Bergström & Hou 2001).

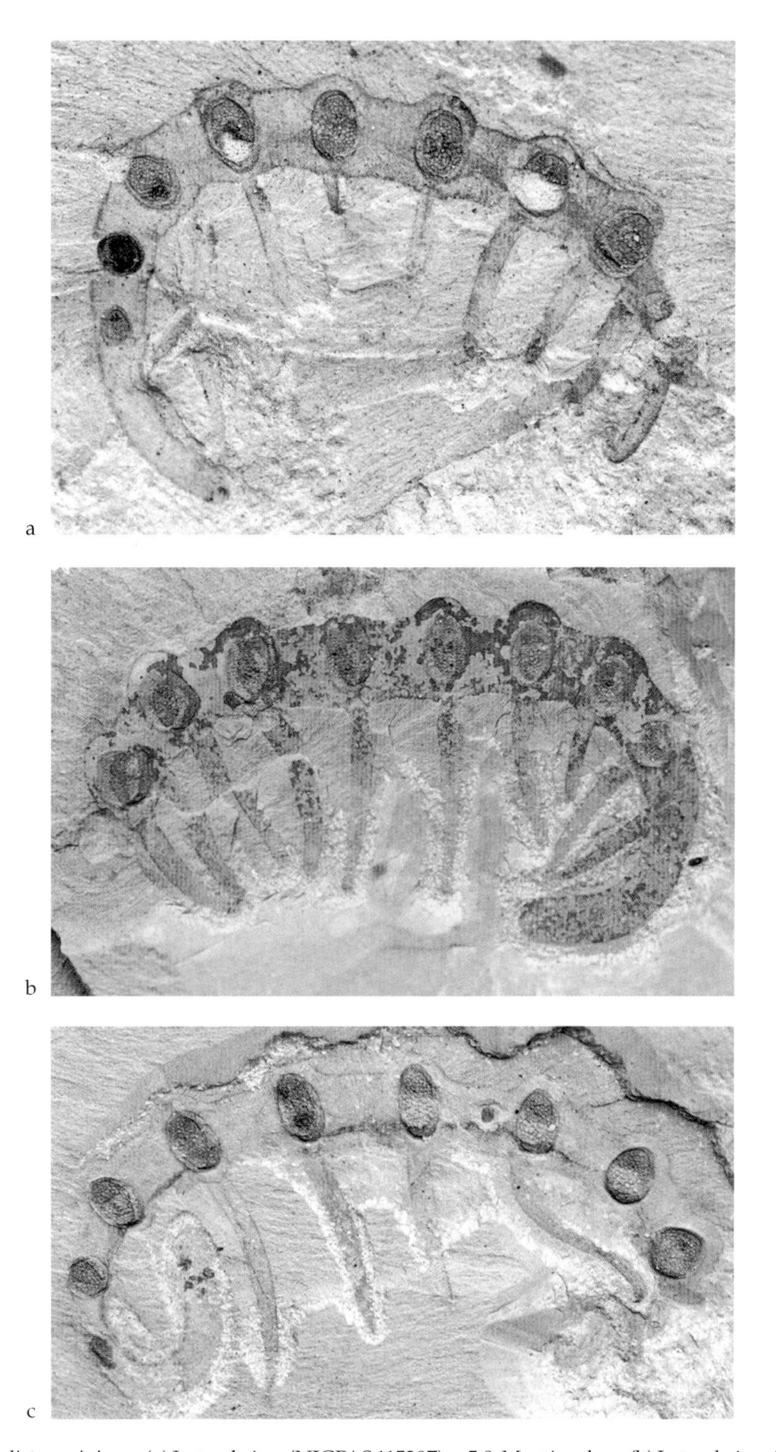

Figure 14.9 *Microdictyon sinicum*. (a) Lateral view (NIGPAS 115297), ×5.8; Maotianshan. (b) Lateral view (RCCBYU 10249), ×4.1; Maotianshan. (c) Lateral view (NIGPAS 115298), ×4.7; Maotianshan.

Genus *Onychodictyon* Hou, Ramsköld & Bergström, 1991

Onychodictyon ferox Hou, Ramsköld & Bergström, 1991

This is one of the more striking of the Chengjiang lobopodians. At least 15 specimens are known. Like the other lobopodians, *Onychodictyon* was mostly soft-bodied or lightly sclerotized, and its fossil remains are now in the form of largely flattened impressions.

This relatively large lobopodian reaches up to about 70 mm in length and 5 mm in width. It has a sclerotized head shield, of poorly-known extent, a feature originally tentatively identified as a jaw structure. Antennae, provisionally identified in early descriptions and reconstructions (Ramsköld & Hou 1991, Hou & Bergström 1995), are now considered to be absent (Ramsköld & Chen 1998). On the trunk there are ten paired scleritic plates. Each plate shows fine reticulate granulation, has an outer ridge and bears a sharp, dorsally directed spine. The plate shape varies from subrectangular posteriorly to more elongate and subrounded anteriorly. Beneath each plate pair there is a pair of legs, and posteriorly there is an eleventh leg pair for which a corresponding plate pair has not been identified. Each of these legs is annulated and bears small appendicules and two large, curved claws. Anterior to the leg pair that is associated with the first plate pair, there is another pair of annulated, claw-bearing legs. These two most anterior leg pairs appear to be somewhat shorter than the others. For much of the trunk there are five annuli between each of the plate/leg pairs, but the trunk area between leg pairs 10 and 11 possibly has just three. On each trunk annulus there is a row of long appendicules symmetrically arranged either side of the mid-line yet apparently being offset longitudinally between successive annuli. The trunk also has very long, clearly annulated appendicules ventrally; posteriorly it does not extend beyond the last (eleventh) leg pair. A linear gut trace is present, as are centrally positioned lines of uncertain nature along each leg.

Onychodictyon contains just *O. ferox*. It has been placed closest to the Mid-Cambrian *Aysheaia* from the Burgess Shale, and also in a larger group with the Early Cambrian *Luolishania* and *Xenusion* from China and Germany, respectively (Ramsköld & Chen 1998).

The claws of *Onychodictyon* clearly had an attachment function, endorsing the notion that it was more adapted to crawling on other organisms than on a muddy sea-bottom. The last leg apparently has anteriorly pointing claws, that were possibly used as "anchors".

O. ferox is known only from the Lower Cambrian of Yunnan Province.

Key References Hou *et al.* 1991, Ramsköld & Hou 1991, Ramsköld 1992, Hou & Bergström 1995, Ramsköld & Chen 1998, Bergström & Hou 2001.

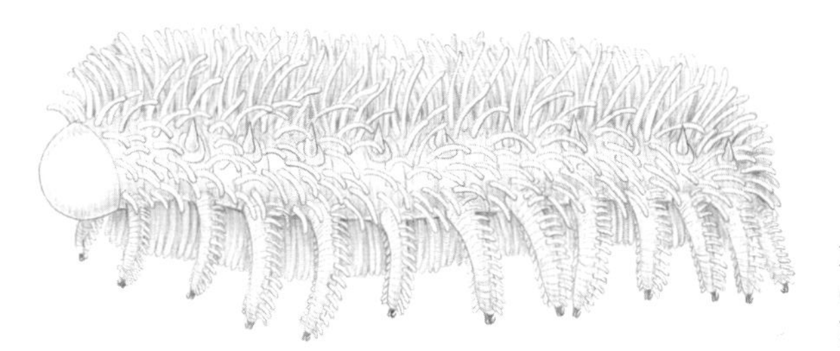

Figure 14.10 Reconstruction of *Onychodictyon ferox* (modified from Bergström & Hou 2001).

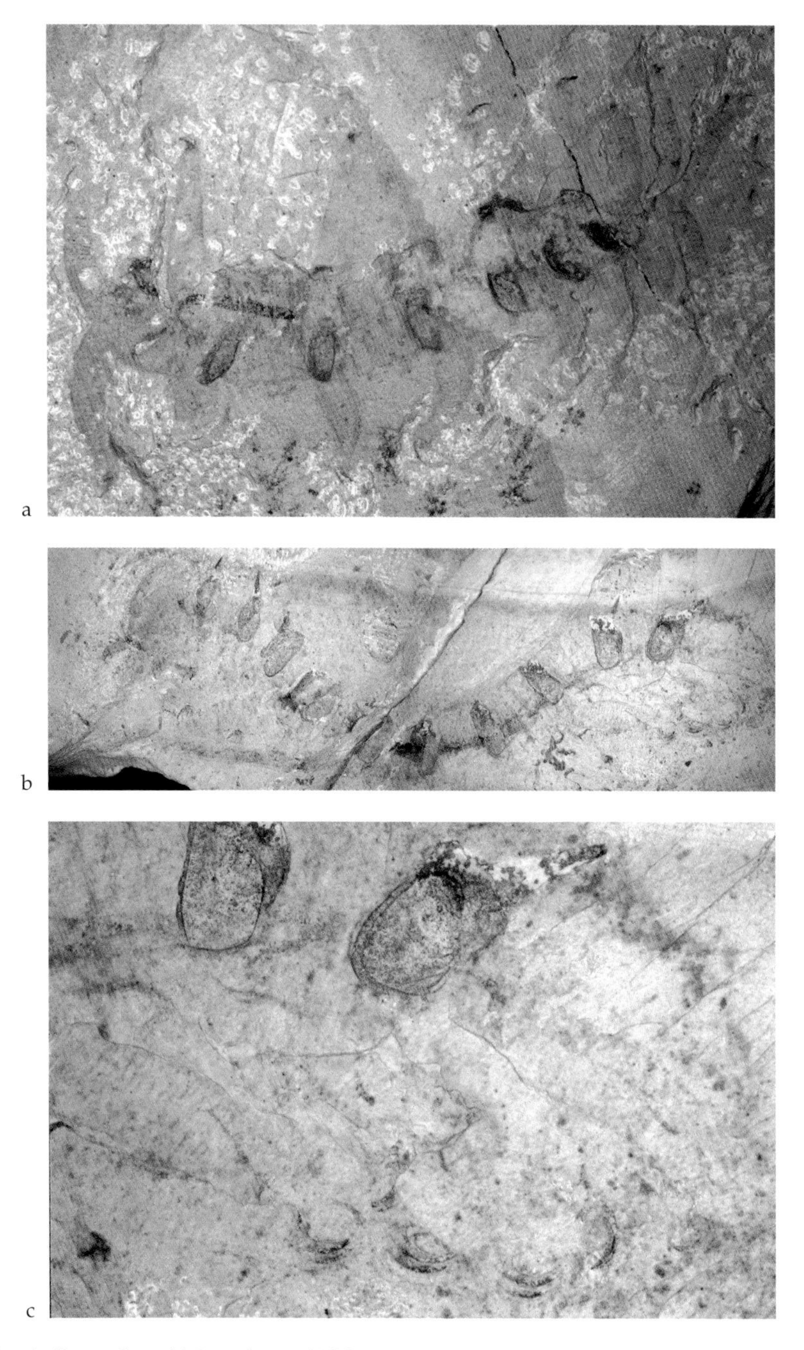

Figure 14.11 *Onychodictyon ferox*. (a) Dorsal view (RCCBYU 10250), × 5.5; Maotianshan. (b) Lateral view (NIGPAS 115271), × 1.9; Maotianshan. (c) Detail of posterior part of trunk, appendages and sclerites (NIGPAS 115271), × 6.6.

15 ANOMALOCARIDIDAE (PHYLUM UNCERTAIN)

These striking Cambrian animals are known from North America, Europe, Australia, and China. They are among the earliest known carnivores and some species reach a size suggested to be well over 1 m. Their morphology recalls features of several phyla, including worm groups, lobopodians and arthropods. They have been regarded as related to one of these groups, or as arthropods, or as forming an unrelated group (see Briggs 1994, Chen *et al.* 1994, Budd 1996, Collins 1996, Wills *et al.* 1998). Possible affinities to the Kinorhyncha, known today from about 150 tiny marine species, have also been suggested (Hou *et al.* 1995). One of the Chengjiang anomalocaridids, *Parapeytoia*, resolves basal to a range of upper stem-group euarthropods (Budd 2002).

Characteristic features of the anomalocaridids include a pair of grasping appendages and a ring of hard plates, armed with projections, surrounding the mouth. Their exoskeleton was probably variably sclerotized. A series of large flaps extends laterally on each side of the body. These were originally thought of as lateral fins, and in other reports they are regarded as ventral appendages. They have been reconstructed as moving either by a wave motion (Whittington & Briggs 1985, Gould 1989, Briggs 1994) or as rowing back and forth to facilitate propulsion (Hou *et al.* 1995). A multisegmented appendage, which may have functioned as a leg, occurs below each lateral flap in some species (Bergström 1986, 1987, Hou *et al.* 1995).

The dorsal side of anomalocaridids has been reconstructed by some authors as more or less smooth except for the presence of the eyes. Several species from Chengjiang and the Burgess Shale show segmentally arranged bands of linear structures that were originally interpreted as lateral gills (Whittington & Briggs 1985). These bands extend over the midline and they have been alternatively interpreted as transverse sets of spines or scales covering the dorsal side of the body (Hou *et al.* 1995). Some species have two long furcal spines, the three segments just anterior of which carry flaps that are notably longer than those further forward.

Four anomalocaridid genera are known from the Chengjiang biota. *Anomalocaris* is also reported from the Canglangpu Formation in Yunnan Province (Zhang *et al.* 2001).

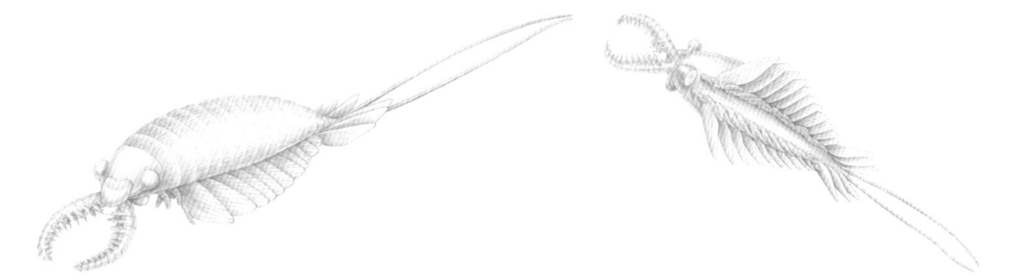

Figure 15.1 Reconstruction of *Anomalocaris saron* in anterolateral and ventral views (based on material of Chen *et al.* 1994 and Hou *et al.* 1995). The presence of the "walking" leg branch of the appendages is based on evidence from other genera.

Genus *Anomalocaris* Whiteaves, 1892

Anomalocaris saron Hou, Bergström & Ahlberg, 1995

This is a relatively common anomalocaridid species, known mostly from about 20 isolated grasping appendages. The maximum size of the animal is unknown, but the incompletely preserved grasping appendages of one particularly large specimen are estimated to have been at least 20 cm long. Like other anomalocaridids, the skin was fairly weakly cuticularized and is therefore as a rule poorly preserved.

The body is fairly slender; excluding grasping appendages and furca its length is about 1.5 times its width including appendages. There is a pair of large, stalked eyes. Each of the pair of massive grasping appendages consists of many short podomeres, most of which have a pair of well-developed multispinose projections along the inner side; a more proximal podomere may carry a small spine. Behind the grasping appendages and the circular mouth is the area of the body from which only simple flaps have been described, but more complex appendages have been reported from this region in *Parapeytoia*. There are 11 pairs of large, overlapping triangular flaps. The upper (or anterior) half of each flap is striated. According to one interpretation, the lower (or posterior) edge of the flaps probably had a clumsy "walking" leg branch, as seen in *Parapeytoia*. Behind these appendages there are an additional three pairs of longer flaps and one pair of long, slender furcal rami.

The spiniferous grasping appendages suggest that this animal was carnivorous. The large appendage flaps would probably have given the animal good swimming ability.

This species is closely related to the type species of the genus, *Anomalocaris canadensis* Whiteaves, 1892 from the Middle Cambrian of British Columbia. *A. saron* is known only from the Chengjiang fauna.

Key References Briggs 1994, Chen *et al.* 1994, Hou *et al.* 1995, Chen & Zhou 1997, Hou *et al.* 1999.

a

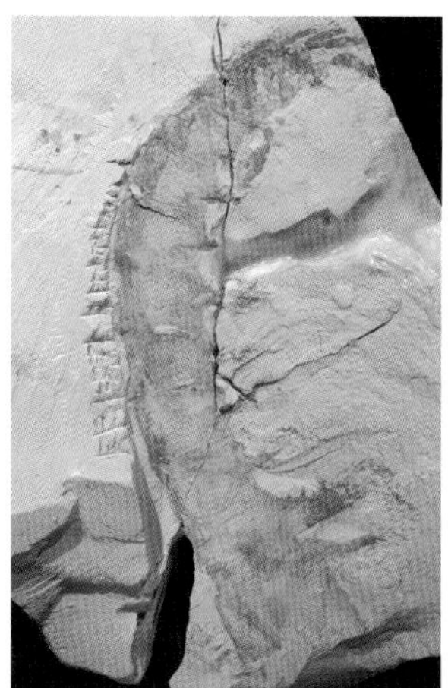
b

Figure 15.2 (a) *Anomalocaris saron*. A disarticulated specimen with grasping appendages (RCCBYU 10314), × 1.5; Mafang. (b) *Anomalocaris* aff. *saron*. A grasping appendage (RCCBYU 10251), × 0.6; Ercaicun.

Genus *Amplectobelua* Hou, Bergström & Ahlberg, 1995

Amplectobelua symbrachiata Hou, Bergström & Ahlberg, 1995

This relatively common anomalocaridid species is typically poorly preserved. Most individuals are represented only by the remains of the grasping appendages, of which about 30 isolated specimens are known.

Some isolated grasping appendages are at least 14 cm long, which gives an idea of the overall large size of the animal. The width of this species, including appendages, is estimated to be virtually the same as its length excluding the terminal furca. *A. symbrachiata* is relatively wider than its Chengjiang associate *Anomalocaris saron*. The paired eyes are large. As in other anomalocaridids, most of the back of the animal is covered by fine transverse sets of lines, which according to one interpretation, represent long slender scales. The grasping appendages differ from those of *A. saron* in having simple, unbranched spines of varying length on 13 podomeres. As in *A. saron*, these projections are aligned in two rows. A spine on the third spiniferous podomere from the proximal end is notably long. The distal end of the grasping appendage has three well-developed, curved spines. The more posterior appendages are known virtually only from their large basal flap. It is not known whether there are elongated posterior appendages as in *Anomalocaris*, but the terminal pair of furcal rami look similar.

The spiny grasping appendages indicate that this species was carnivorous like other anomalocaridids. The overall similarity in the grasping appendages between this species and *Anomalocaris saron* hints that the genera are closely related. The main known difference is that the spines of the appendages are simple in *Amplectobelua* but branched in *Anomalocaris*.

Amplectobelua is known from a single species that is recorded only from the Chengjiang biota.

Key References Chen *et al.* 1994, Hou *et al.* 1995, Chen & Zhou 1997, Hou *et al.* 1999.

a

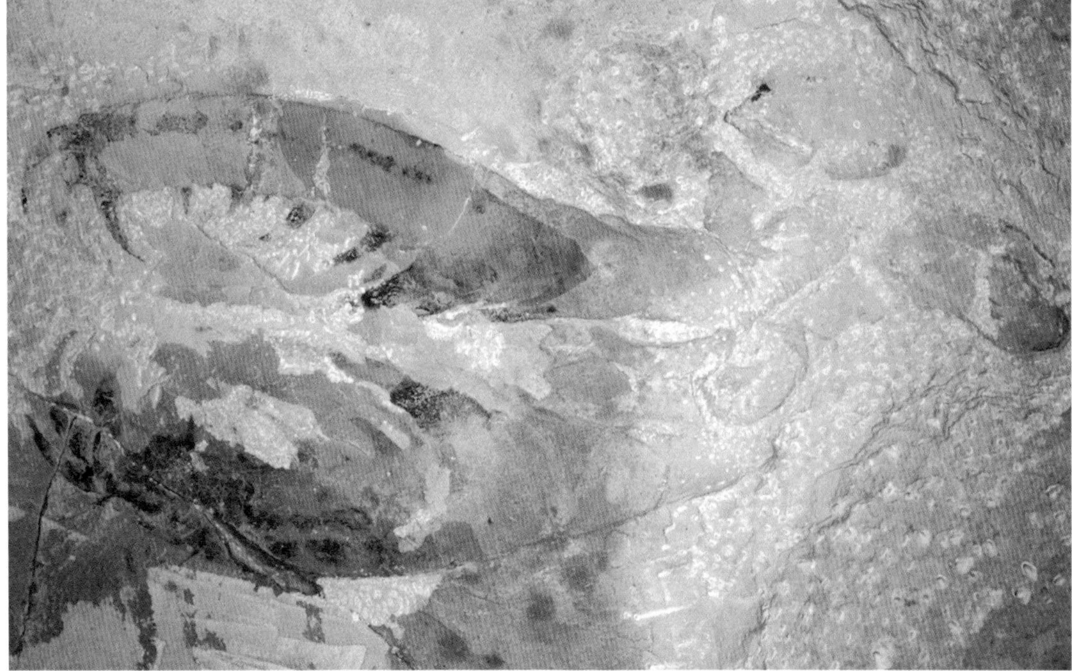

b

Figure 15.3 *Amplectobelua symbrachiata*. (a) An isolated grasping appendage (RCCBYU 10252), ×1.4; Ma'anshan. (b) A pair of isolated grasping appendages (NIGPAS 115346), ×3.4; Maotianshan.

Genus *Cucumericrus* Hou, Bergström & Ahlberg, 1995

Cucumericrus decoratus Hou, Bergström & Ahlberg, 1995

This is a rare species, of which only limbs and fragments of cuticle are preserved. The entire length of the animal is unknown.

A trunk appendage consists of a large, striated triangular flap and a "walking" leg branch. Proximally, where these rami appear to be confluent, there are several small, medially facing endites. The "walking" leg, which is estimated to be at least 5.5 cm long, has well-defined podomeres only in its distal part. The dorsal surface of the animal is traversed by shallow furrows and appears to have been a fairly soft and pliable cuticle, rather than forming a stiff exoskeleton.

The leg morphology of this species demonstrates a state of incipient segmentation, a feature that also might be expected to occur in the earliest arthropods. On the other hand, *Cucumericrus* and other anomalocaridids are decidedly advanced in the development of the mouth and grasping appendages, whereas many early arthropods were unspecialized in those respects. The arthropod-like features of anomalocaridids could be interpreted as a case of convergence. An alternative interpretation, having anomalocaridids as stem-group arthropods, has also been suggested (e.g. Budd 1996, 2002, Wills *et al.* 1998).

Although there is only limited knowledge of the morphology of *C. decoratus*, like other anomalocaridids it is likely to have been a swimmer and a predator. It is the only known species of its genus and is not recorded outside the Chengjiang fauna.

Key References Hou *et al.* 1995, Hou *et al.* 1999.

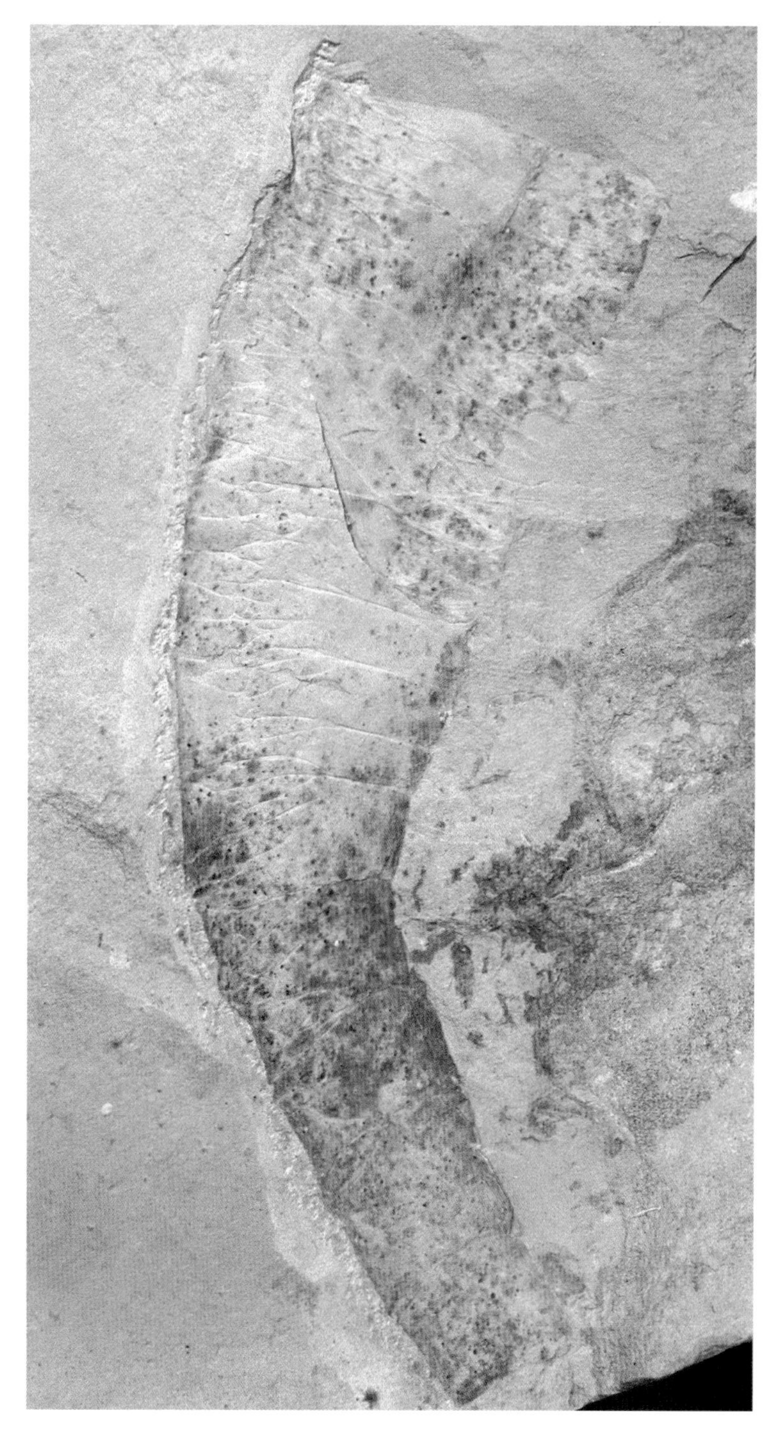

Figure 15.4 *Cucumericrus decoratus*. An isolated "walking" leg (NIGPAS 115351), ×3.0; Maotianshan.

Genus *Parapeytoia* Hou, Bergström & Ahlberg, 1995

Parapeytoia yunnanensis Hou, Bergström & Ahlberg, 1995

This is a rare anomalocaridid and may have been a fairly small species; the grasping appendage of one specimen is estimated to be about 5 cm long.

P. yunnanensis differs notably from species of the anomalocaridids *Anomalocaris* and *Amplectobelua* by the morphology of its grasping appendage, which has only a few podomeres, three of which each have long, serrated finger-like extensions. These extensions and the long, pointed, distal-most podomere together form an unusual tong-like serial set. The succeeding two pairs of appendages, near the mouth, are quite small and appear to have assisted in the manipulation of the food. Behind, in the trunk, follows the typical anomalocaridid set of wide, striped flaps extending laterally. Along the lower medial edge of each flap there are five endites; adjacent to these, attached to the flap, is a multisegmented free leg. There is an indication of the presence of a pair of long trailing spines, as in *Anomalocaris*. The mouth ring is also similar to that of *Anomalocaris*. The paired eyes in *Parapeytoia* are probably positioned further back than in *Anomalocaris*. According to one interpretation, the dorsal side of *P. yunnanensis* has transverse sets of lanceolate scales.

The spiny grasping appendages indicate that this species was carnivorous. The grasping appendages are similar in several respects to those of *Peytoia* from the Middle Cambrian Burgess Shale, indicating that they are possibly closely related.

P. yunnanensis is known only from the Chengjiang fauna.

Key References Hou *et al.* 1995, Hou *et al.* 1999.

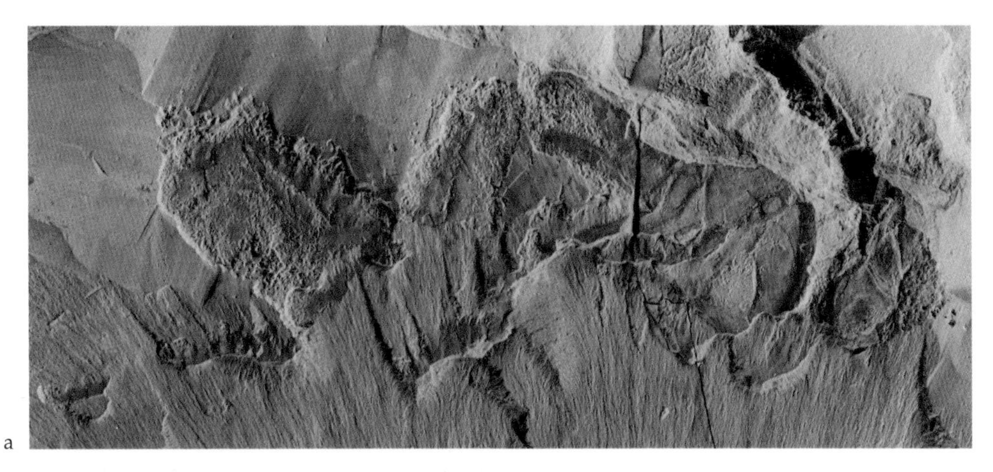

Figure 15.5 *Parapeytoia yunnanensis*. (a) Ventral view, part (NIGPAS 115334b), × 1.3. (b) Ventral view, counterpart, showing grasping appendages, mouth, sternites and several other pairs of appendages (NIGPAS 115334a), × 1.6; Maotianshan.

16 PHYLUM ARTHROPODA

Arthropods are an extremely abundant, ecologically diverse and very successful group. A segmented body, a chitinous cuticle forming an external skeleton, jointed limbs, well-developed sensory systems, and a life cycle involving molting are some of the features that characterize the group. The adjoining plates of cuticle (sclerites) that form the exoskeleton consist of dorsal (tergite) and ventral (sternite) parts. The myriapods (millipedes, centipedes) and the hexapods (e.g. insects) have sets of paired limbs, in which there is a single (uniramous) main branch to the limb. Those arthropods with biramous limbs, having inner (endopod) and outer (exopod) branches, include the trilobites, the crustaceans (e.g. shrimps), and the chelicerates. The Chelicerata embraces xiphosurids (e.g. king crabs), eurypterids (sea scorpions), and arachnids (spiders and scorpions). The trilobites, chelicerates, and allies are grouped as the arachnomorphs. Basal to all these there are minor but evolutionary significant groups such as the tardigrades (water bears).

The Recent arthropod fauna is dominated by crustaceans, chelicerates, hexapods, and myriapods. Trilobites (Cambrian-Permian), together with ostracod and other crustacean groups, in which the cuticle is mineralized with calcium carbonate or calcium phosphate, comprise most of the fossil arthropods. Molecular and morphological evidence yield a plethora of models of arthropod phylogeny (see Edgecombe 1998, Fortey & Thomas 1998, Giribet *et al.* 2001, Maas & Waloszek 2001). Most current commentators consider that arthropods are monophyletic.

Arthropods form the major component of the Chengjiang fauna; they are documented in more than 100 species names that represent about 60 biological species. The tiny bradoriids are the most species diverse and abundant group, individuals occurring in thousands (e.g. *Kunmingella*). Fossils interpreted as basal arthropods (e.g. *Fuxianhuia*), chelicerates (e.g. *Parapaleomerus*), and crustaceans (e.g. *Pectocaris* and *Waptia*) have been described from the biota. However, there are contrasting opinions regarding the systematic position of many of the taxa. For example, some analyses resolve naraoiids within the Trilobita (e.g. Wills *et al.* 1998), others place them outside (e.g. Bergström & Hou 1998, Edgecombe & Ramsköld 1999b). Superficial similarity can be deceiving, and soft part morphology often reveals an alternative affiliation. A few of the Chengjiang species may be related to crustaceans, but many forms probably belong among a "bush" of evolutionary off-shoots from the arthropod stem. In one analysis several such problematic Cambrian arthropods, known from the Chengjiang and Burgess Shale biotas, resolve as a clade close to crown-group euarthropods (Budd 2002). Some of them have developed large (so-called "great") appendages behind a set of simple antennae (e.g. *Fortiforceps*). Several Chengjiang arthropods have a "carapace", which in some species is bivalved. In some forms an abdomen and tail extend well beyond the carapace (e.g. *Clypecaris*), in others (e.g. the bradoriids) the carapace covers all or almost all the entire body.

Many Chengjiang arthropods have simple legs. With this simplicity goes a lack of specialized mouthparts in many species, and a gut containing what looks like sediment, implying deposit-feeding activities. Predatorial and scavenging lifestyles and pelagic swimmers are also represented.

Genus *Urokodia* Hou, Chen & Lu, 1989

Urokodia aequalis Hou, Chen & Lu, 1989

U. aequalis was described from about 15 specimens from the Chengjiang area. It is an elongate species, some 35 mm long, with a shield at both ends and a myriapod-like trunk consisting of about 14 similar tergites. The head shield bears a pair of anterior spines and three pairs of lateral spines. The tail shield is almost equal in size and was originally considered to be morphologically similar to the head shield. A subsequently found, complete *Urokodia* specimen from Anning, Yunnan Province, has two large paired spines and many small ones on the tail (Zhang *et al.* 2002), but it is not certain that it is the same species. The only known soft part of *Urokodia* is a possible stout antenna.

It is difficult to resolve the affinities of *U. aequalis*, or of similar forms such as *Mollisonia* Walcott, 1912, especially because appendages are not preserved. Its form indicates that *U. aequalis* probably lived on the sea-bottom. Since its appendages are unknown, so too is its mode of feeding. This species is only known from the Chengjiang biota.

Key References Hou *et al.* 1989, Hou *et al.* 1999, Zhang *et al.* 2002.

a

b

Figure 16.1 *Urokodia aequalis.* (a) Dorsal view of specimen with head shield (NIGPAS 108312), × 3.1; Jianbaobaoshan, near Dapotou. (b) Lateral view of trunk (RCCBYU 10253), × 3.0; Jianbaobaoshan, near Dapotou.

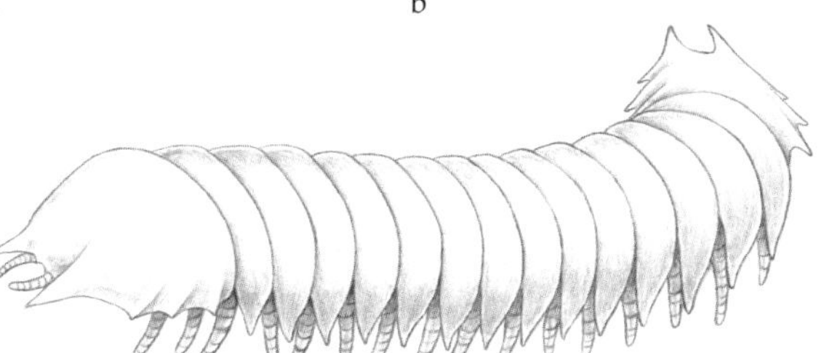

Figure 16.2 Reconstruction of *Urokodia aequalis* (the morphology of the appendages is hypothetical).

Genus *Fuxianhuia* Hou, 1987

Fuxianhuia protensa Hou, 1987

This species is known from hundreds of specimens, the largest of which are about 11 cm long. They are usually fairly complete and preserved flattened.

The "head shield" and broad anterior part of the trunk are succeeded by a narrower, elongate abdomen. The "shield" covers the head and extends back as a (carapace) fold over three additional segments. The head carries a pair of short, annulate antennae in front of a pair of curved, uniramous grasping appendages each consisting of about eight podomeres. There is evidence of two additional pairs of appendages in the head, each with an endopod and a flat exopod (Hou & Bergström 1997). A pair of eyes project anteriorly to the head. Behind the head there are about 31 trunk tergites: the first three thoracic tergites are followed by 13–15 tergites forming the main, broad part of the thorax; the abdomen has about 13 tergites and a terminal spine flanked by a pair of lateral spines. The biramous legs of the thorax have a sturdy inner, "walking" branch consisting of about 25 similarly shaped podomeres and a thin, oval-shaped exopod with smooth margins. The legs show no terminal claws or spines. The leg morphology thus appears primitive, showing little sophistication except for the addition of a leaf-like outer branch. Although some 16–18 trunk tergites carry legs, there appear to be as many as 35–45 pairs of legs in total, with 2–4 pairs of legs for each tergite of the thorax (Hou & Bergström 1997; compare Chen *et al.* 1995b). Thus, repetition on the dorsal side does not seem to match that on the ventral side.

The systematic position of *Fuxianhuia* is controversial. It has been regarded as a basal euarthropod (Chen *et al.* 1995b, Edgecombe & Ramsköld 1999a) and a possible early chelicerate (Wills 1996). Another analysis resolves this genus as an upper stem-group euarthropod close to the Chengjiang genus *Pectocaris* (Budd 2002; see also Maas & Waloszek 2001). Hou & Bergström (1997) assigned *Fuxianhuia* to a new family and a new superclass Proschizoramia, which they characterized as a group at an early stage in the evolution of arthropods with biramous limbs. Certainly, not all the characters of *Fuxianhuia* are primitive; advanced features include the development of grasping appendages in a defined head region.

Although its overall appearance suggests that this was a benthic animal, its crowded, somewhat clumsy-looking legs hint that it was a slow mover. The grasping pair of appendages in the head indicates that it may have been a carnivore. The gut of some individuals is stuffed with mud and the possible remains of small animal prey that may have been engulfed indiscriminately together with the sediment (Hou & Bergström 1997).

F. protensa is unknown outside the Chengjiang biota.

Key References Hou 1987b, Chen *et al.* 1995b, Wills 1996, Chen & Zhou 1997, Hou & Bergström 1997, Bergström & Hou 1998, Hou *et al.* 1999, Budd 2002.

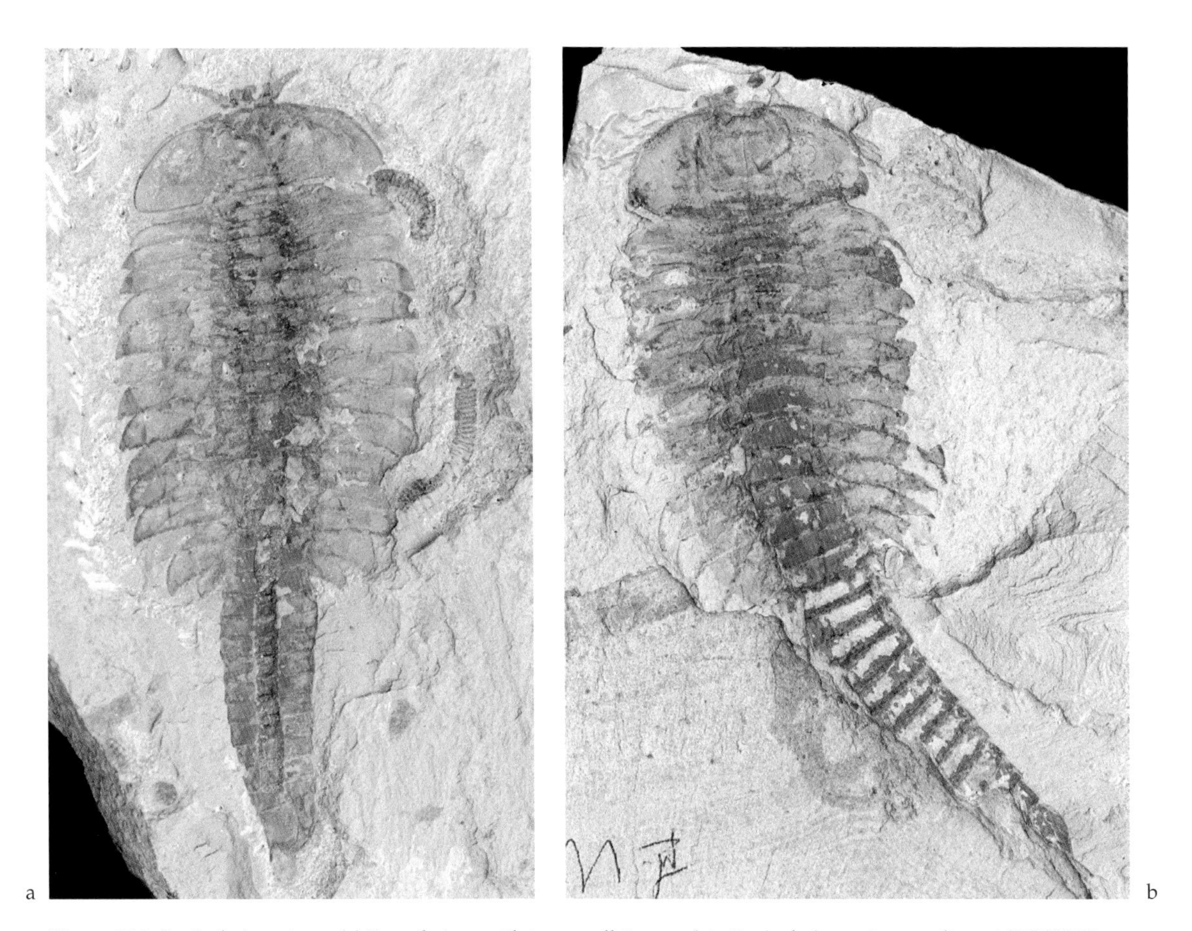

Figure 16.3 *Fuxianhuia protensa*. (a) Dorsal view, with two small, incomplete *Fuxianhuia* specimens adjacent (RCCBYU 10254), ×1.7; Mafang. (b) Dorsal view (RCCBYU 10255), ×1.5; Mafang.

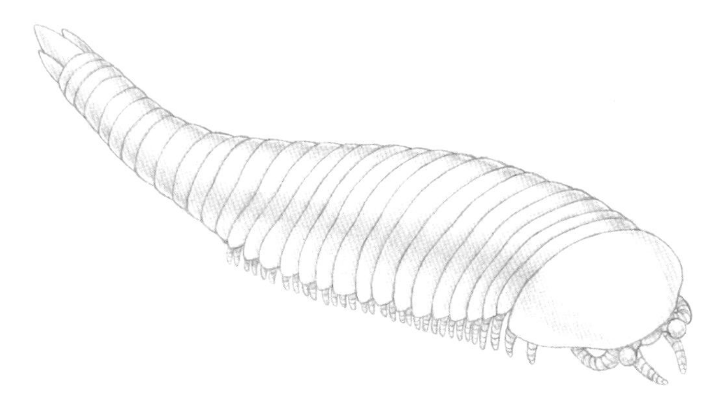

Figure 16.4 Reconstruction of *Fuxianhuia protensa*.

Genus *Chengjiangocaris* Hou & Bergström, 1991

Chengjiangocaris longiformis Hou & Bergström, 1991

This species is known from only a few specimens. Both dorsoventral compression and flattening from the sides are represented.

C. longiformis does not exceed 10 cm in length. An indistinct trilobation is seen along the entire trunk. Anteriorly, in the holotype, there is a series of about five short tergites (presumed thorax), but the head area is missing. Other, new material displays antennae and stalked eyes. Behind the five thoracic tergites the trunk has another 17 tergites (abdomen) plus a triangular-shaped terminal element. In contrast to *Fuxianhuia protensa*, to which it shows several similarities, the trunk lacks obvious differentiation into a broad middle part and a slender posterior part. The endopods in the trunk are simple, each consisting of about 20 uniform podomeres and a small conical end piece. The exopod is a simple rounded flap, lacking setae or bristles. The appendages are much more closely spaced than the trunk tergites (Hou & Bergström 1997), resulting in a "segmental mismatch" of the type known from *Fuxianhuia protensa*, though the exact pattern of the mismatch in *C. longiformis* is not known.

It has been suggested that *Chengjiangocaris* is related to *Sanctacaris* Briggs & Collins, 1988 and emeraldellids of the Burgess Shale (Delle Cave & Simonetta 1991). This has been disputed by Hou & Bergström (1997) who considered that an affinity with *Fuxianhuia* is much more likely and reconstructed the head morphology of *C. longiformis* based on *Fuxianhuia*. These supposedly primitive arthropods with apparently segmental mismatch and simple legs may hint at characters of the original arthropod. *Chengjiangocaris* is the type and only genus of the Family Chengjiangocarididae. The Chengjiang taxon *Cambrofengia yunnanensis* may be based on detached appendages of *C. longiformis.*

The habitat and benthic mode of life of *C. longiformis* may have been similar to that of the related *Fuxianhuia*. *C. longiformis* is known only from the Chengjiang biota.

Key References Hou & Bergström 1991, Hou & Bergshröm 1997, Hou *et al.* 1999.

a

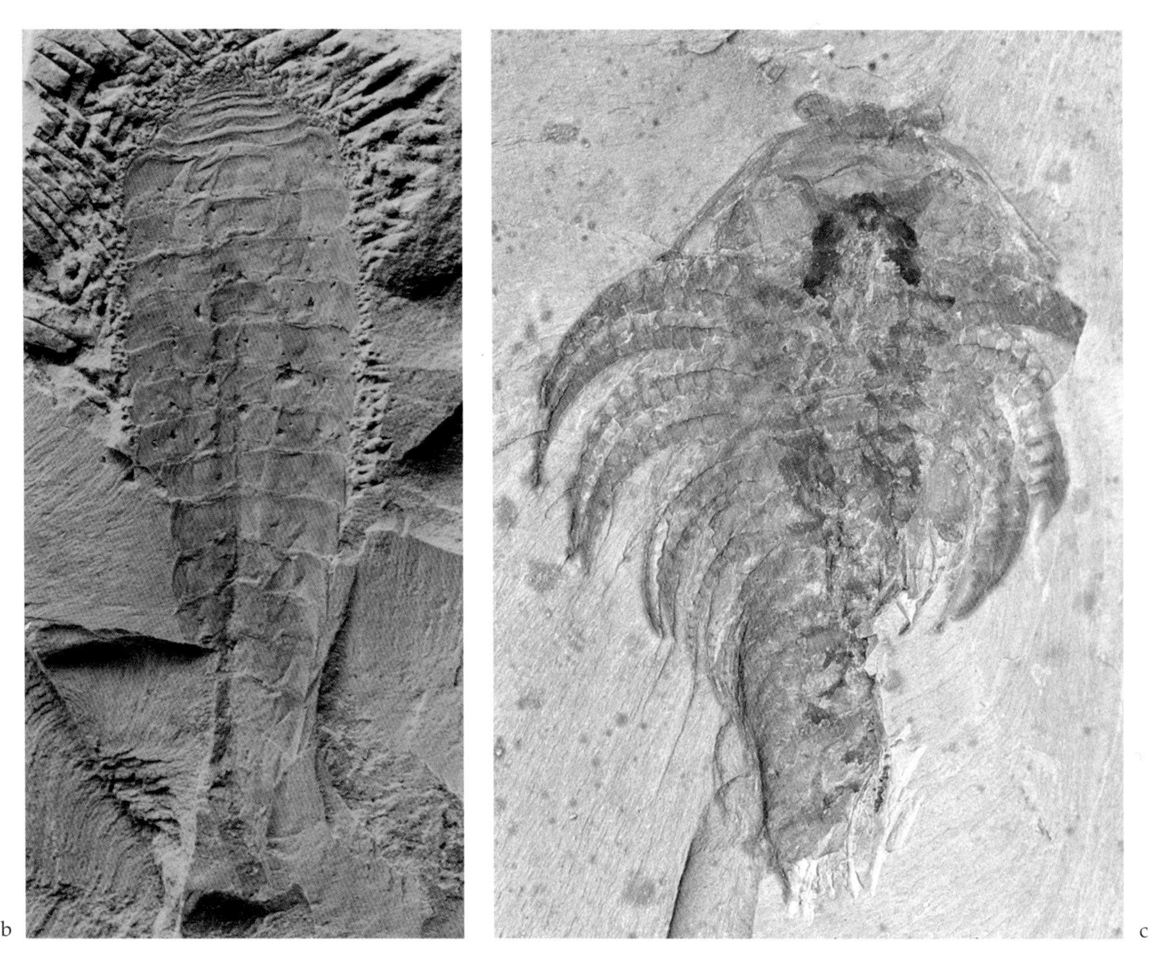

Figure 16.5 *Chengjiangocaris longiformis*. (a) Anterolateral view of strongly flexed specimen (NIGPAS 115359), ×1.7; Xiaolantian. (b) Dorsal view, counterpart (NIGPAS 110837), ×1.6; Maotianshan. (c) Ventral view (RCCBYU 10256), ×2.4; Xiaolantian.

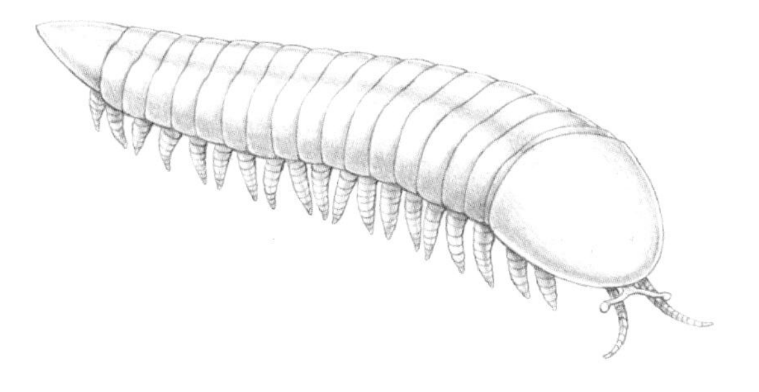

Figure 16.6 Reconstruction of *Chengjiangocaris longiformis*.

Genus *Pisinnocaris* Hou & Bergström, 1998

Pisinnocaris subconigera Hou & Bergström, 1998

P. subconigera is a rare species, known from only a few incomplete dorsoventrally flattened specimens, the largest of which is about 11 mm long.

The head shield forms the broadest part of the animal and has a rounded outline. A pair of eyes and simple paired antennae are preserved at the anterior margin of the head, where there is also a rostral plate. Behind the head shield there is a regularly tapering trunk with at least ten overlapping tergites. No appendages are known, but there are some color patterns indicating a maximum of three pairs of muscle scars (and therefore possibly appendages) corresponding to a single tergite. The relatively long tergites in the trunk may reflect the possession of more than one pair of ventral appendages per tergite. In life, the animal appears to have been highly vaulted in cross-section.

From what is known of its morphology, particularly the nature of the head shield and its soft parts, and the possible mismatch of appendages and segments in the trunk, *Pisinnocaris* may have affinities with *Fuxianhuia* and *Chengjiangocaris* of the Chengjiang biota (Hou & Bergström 1998). A segmental mismatch is also seen in the Chengjiang genera *Xandarella* and *Cindarella*, but these animals are different in other respects. The Chengjiang species *Jianshania furcatus* appears to be a possible synonym of *P. subconigera*.

In the absence of knowledge of the structure of its limbs, attempts at determining the mode of life of *P. subconigera* are difficult. As its body is fairly broad it may have been a bottom dweller. The species is known only from the Chengjiang fauna.

Key References Hou & Bergström 1998, Hou *et al.* 1999.

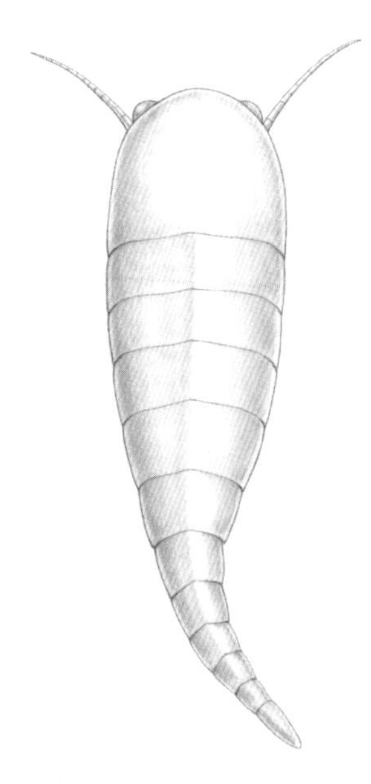

Figure 16.7 Reconstruction of *Pisinnocaris subconigera*. The morphology of the tail piece is hypothetical.

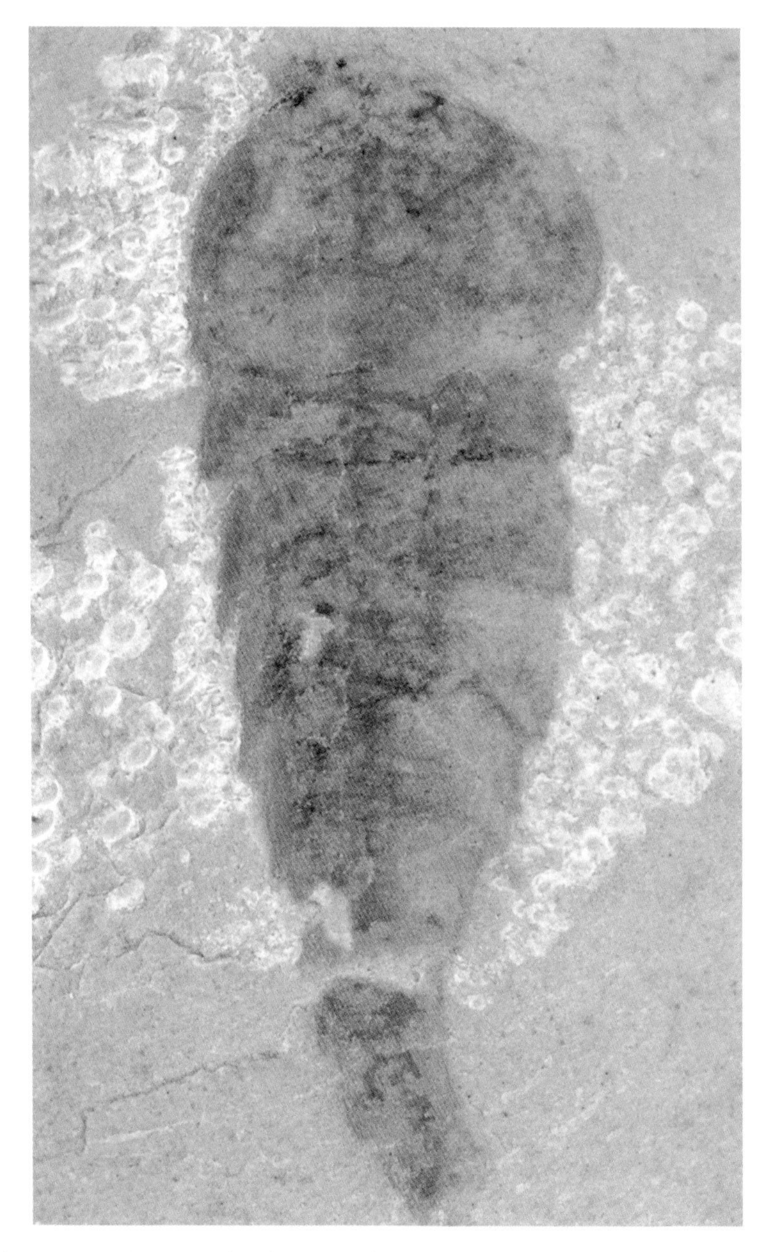

Figure 16.8 *Pisinnocaris subconigera*. Dorsal view (NIGPAS 115417b), × 17.6; Maotianshan.

Genus *Dongshanocaris* Hou, Bergström, Wang, Feng & Chen, 1999

Dongshanocaris foliiformis (Hou & Bergström, 1998)

D. foliiformis is known from a single fossil almost 2 cm long. Behind the poorly-preserved head the specimen is fairly complete, but details of the morphology are difficult to discern.

The head seems to have three or four segments. Behind, there are 26 tergites each with a single pair of appendages (presumed thorax), and a tail consisting of 12 very tightly spaced segments. The appendages are biramous and are exposed marginal to the trunk. The exopod appears to be a simple flap with a curved outline, but the endopod is too poorly preserved for details to be interpreted.

The systematic position of this taxon is uncertain. Some aspects of the trunk morphology of *Dongshanocaris* recall the Burgess Shale genus *Marrella* Walcott, 1912 and the Devonian Hunsrück Slate genus *Mimetaster* Gürich, 1932 but their appendages are quite different (Hou & Bergström 1998). The simple exopod of the monotypic *Dongshanocaris* is in general comparable to the exopods of *Fuxianhuia* and *Chengjiangocaris*, but there is no other evidence to suggest that these genera are allied.

The mode of life of this species is uncertain. The only known specimen is from Maotianshan.

Key References Hou & Bergström 1998, Hou *et al.* 1999.

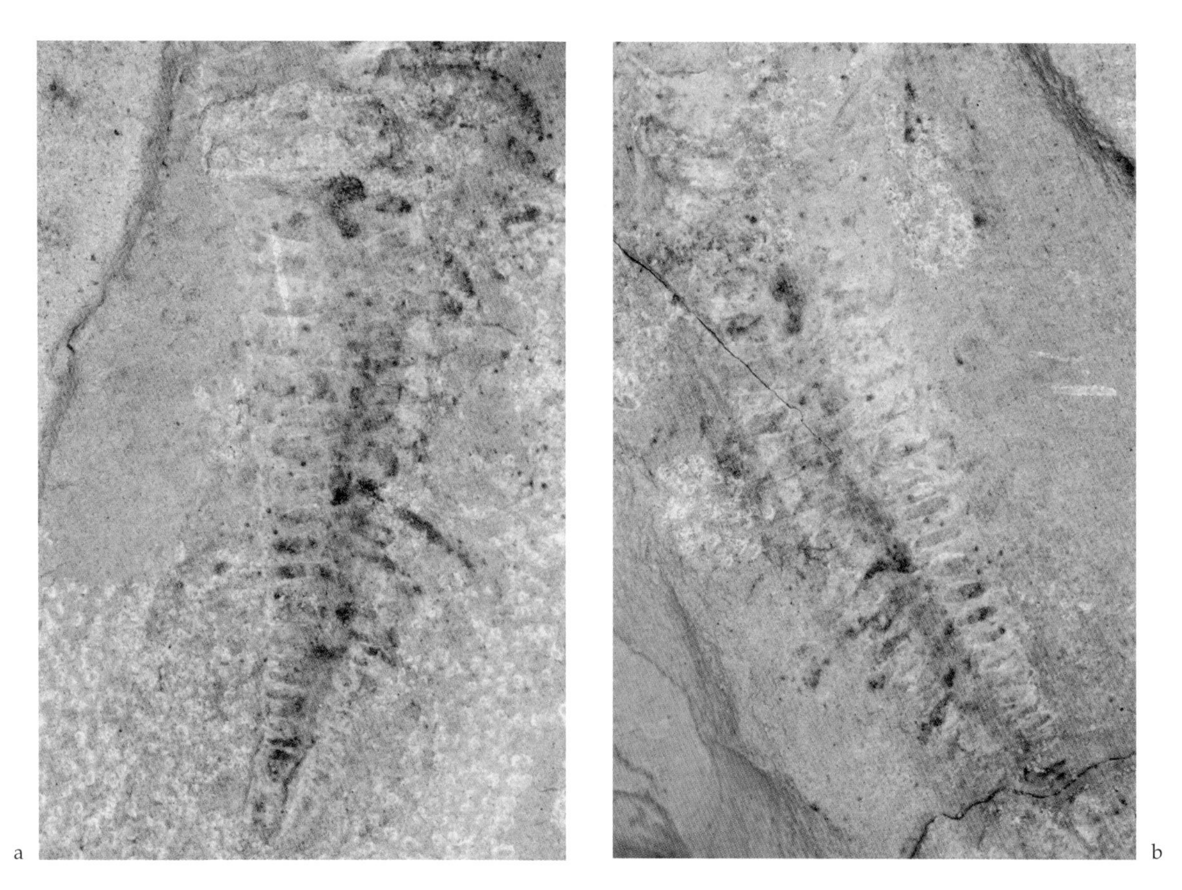

Figure 16.9 *Dongshanocaris foliiformis*. (a) Dorsal view, part (NIGPAS 115419a), ×6.6; Maotianshan. (b) Dorsal view, counterpart (NIGPAS 115419b), ×7.1.

Genus *Canadaspis* Novozhilov *in* Orlov, 1960

Canadaspis laevigata (Hou & Bergström, 1991)

This relatively common species is usually preserved in lateral aspect. It has a shrimp-like appearance overall, due in part to the presence of a large bivalved carapace covering the head and a part of the trunk. Terminally there is an elongate telson flanked by spines. The carapace is less than 2 cm long, and the entire animal is less than 3 cm long.

The head has a pair of stalked eyes and a pair of uniramous antennae that project anteriorly beyond the carapace. There are at least 21 segments between the eyes and the telson, about 19 of which are in the trunk (Hou & Bergström 1997). Behind the antennae there are at least ten pairs of biramous appendages, the first few pairs being in the head. The stout, multisegmented "walking" branch ends distally in a set of spines; the exopod is a large, flat, ovoid blade with a wrinkled appearance in places.

This Chengjiang species is similar to the Burgess Shale type species, *Canadaspis perfecta* (Walcott, 1912), from which it differs in number of segments, and in its less spiny telson and smaller carapace. *Canadaspis* has been interpreted as a crustacean (Briggs 1992, Wills 1998a, Wills *et al.* 1998), but this has been denied on the basis of limb morphology (Dahl 1984, Boxshall 1998, Walossek 1999, Maas & Waloszek 2001). Its limb structure seems to reflect a primitive level of organization, comparable with that of *Fuxianhuia* and its allies (Hou & Bergström 1997). Budd's (2002) analysis of upper stem-group euarthropods indicates that *Canadaspis* is closely related to *Odaraia* and *Branchiocaris*.

The Chengjiang species *Canadaspis eucalla* is thought (Hou *et al.* 1999) to be synonymous with C. *laevigata*. The poorly known *Yiliangocaris ellipticus* from Yunnan Province is possibly another junior synonym.

This species seems to lack appendages specialized for feeding. The common occurrence of what has been identified as silt in the gut indicates that its food supply was probably the organic contents of ingested sediment (Hou & Bergström 1997). Briggs *et al.* (1994) thought that *Canadaspis* used its limbs to churn up sediment in search of small animals and organic particles on which it fed.

Canadaspis is known from the Lower and Middle Cambrian of China and from the Middle Cambrian of North America.

Key References Hou & Bergström 1991, Chen & Zhou 1997, Hou & Bergström 1997, Bergström & Hou 1998, Hou *et al.* 1999, Budd 2002.

Figure 16.10 Reconstruction of *Canadaspis laevigata* (modified from Hou & Bergström 1997).

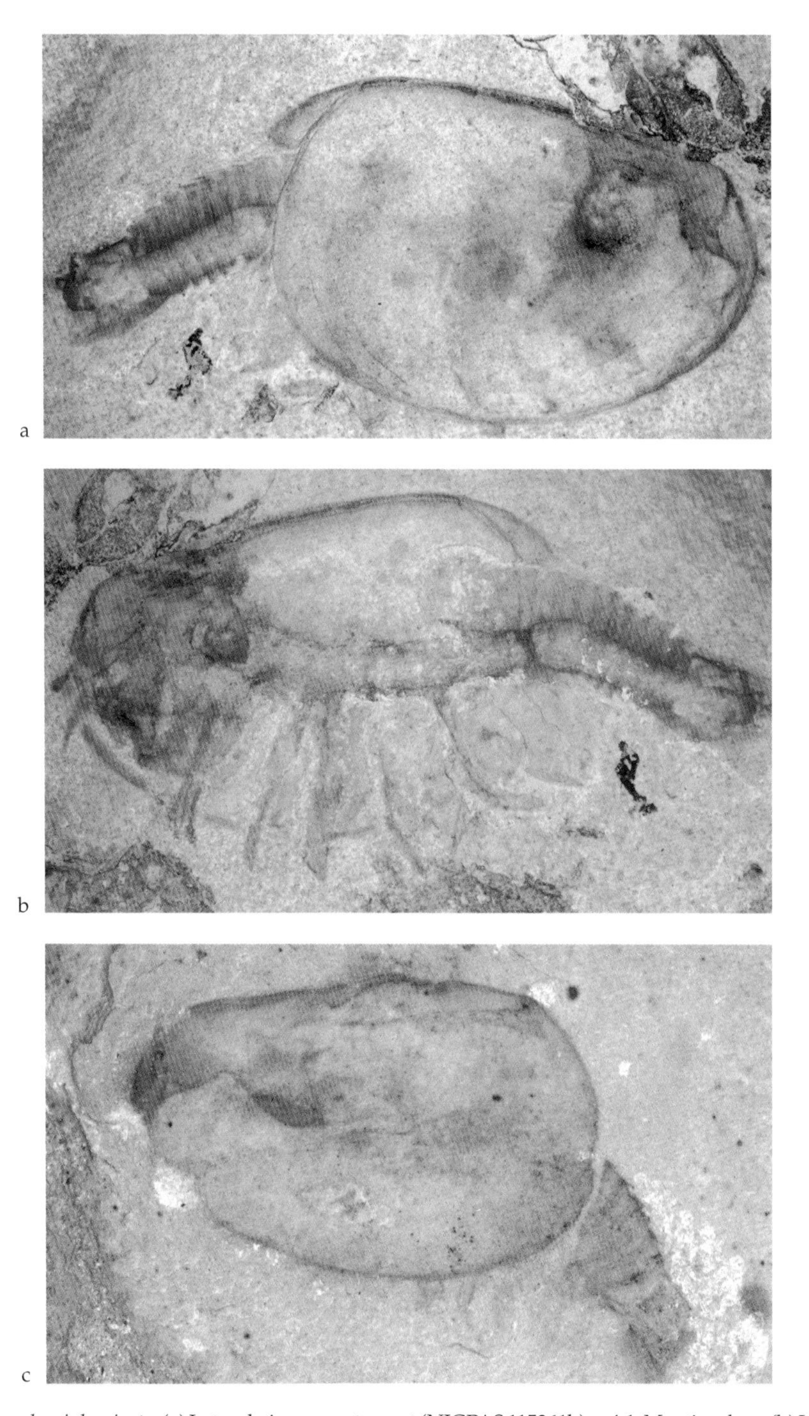

Figure 16.11 *Canadaspis laevigata*. (a) Lateral view, counterpart (NIGPAS 115361b), ×4.1; Maotianshan. (b) Lateral view, part, after preparation (NIGPAS 115361a), ×4.1. (c) Lateral view (RCCBYU 10257), ×6.4; Maotianshan.

Genus *Kunmingella* Huo, 1956

Kunmingella douvillei (Mansuy, 1912)

This bivalved arthropod is the most abundant of the many bradoriid species in the Chengjiang biota. Thousands of its thin, weakly-mineralized valves occur at many horizons, accounting for a high percentage of recovered individuals from the biota. Conjoined valves, preserved in articulated "butterfly" orientation on bedding planes, are common. Soft parts are known from some 45 specimens.

Adult valves are generally smooth, about 5 mm long, with a tapering posterior lobe, an anterodorsal node and a narrow marginal ridge. A pair of short, uniramous, sensory antennae occurs in front of seven pairs of biramous appendages, of which probably three pairs are on the head (Hou *et al.* 1996, Shu *et al.* 1999). The leaf-shaped exopods, fringed with marginal spines, may have had a respiratory function. Each endopod is elongate, consisting of at least five podomeres, and was probably used for walking. The podomeres of the cephalic endopods have a long spine-like extension on their inner margin, giving each of these limb branches an overall rake-like appearance. The endopods of the posterior-most trunk appendage are particularly long and slender. Definition of the trunk end is obscure, but it may be a narrow triangular piece flanked by a pair of furca-like rami. This tail piece and parts of several of the posterior biramous limbs project beyond the posterior margin of the carapace in some specimens.

Bradoriids have traditionally been regarded as ostracod crustaceans. However, the soft parts of *Kunmingella* indicate that bradoriids belong outside the Crustacea *sensu stricto* and are, at most, early derivatives of the stem line Crustacea (Hou *et al.* 1996, Shu *et al.* 1999). Bradoriids are just one of several groups of Cambrian arthropods that, convergently, developed bivalved shells.

K. douvillei probably crawled on, and swam near, the substrate, protected by its carapace. The relatively short antennae may represent a reduction in favor of the development of large eyes, the presence of which is suggested in the prominent anterodorsal node. The occurrence of supposed coprolites rich in *Kunmingella* indicates that bradoriids were a food source for larger predators.

K. douvillei is known only from the Cambrian of China, the country with the richest bradoriid faunas globally (Hou *et al.* 2002b).

Key References Hou *et al.* 1996, Shu *et al.* 1999, Hou *et al.* 2002b.

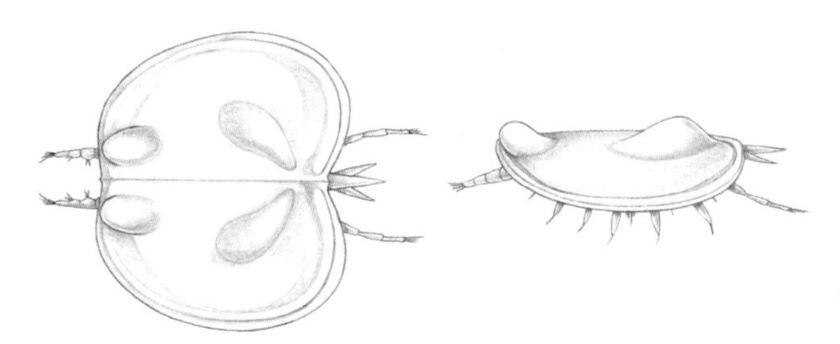

Figure 16.12 Reconstruction of *Kunmingella douvillei* in dorsal and lateral views (after Shu *et al.* 1999; based in part on evidence in Hou *et al.* 1996).

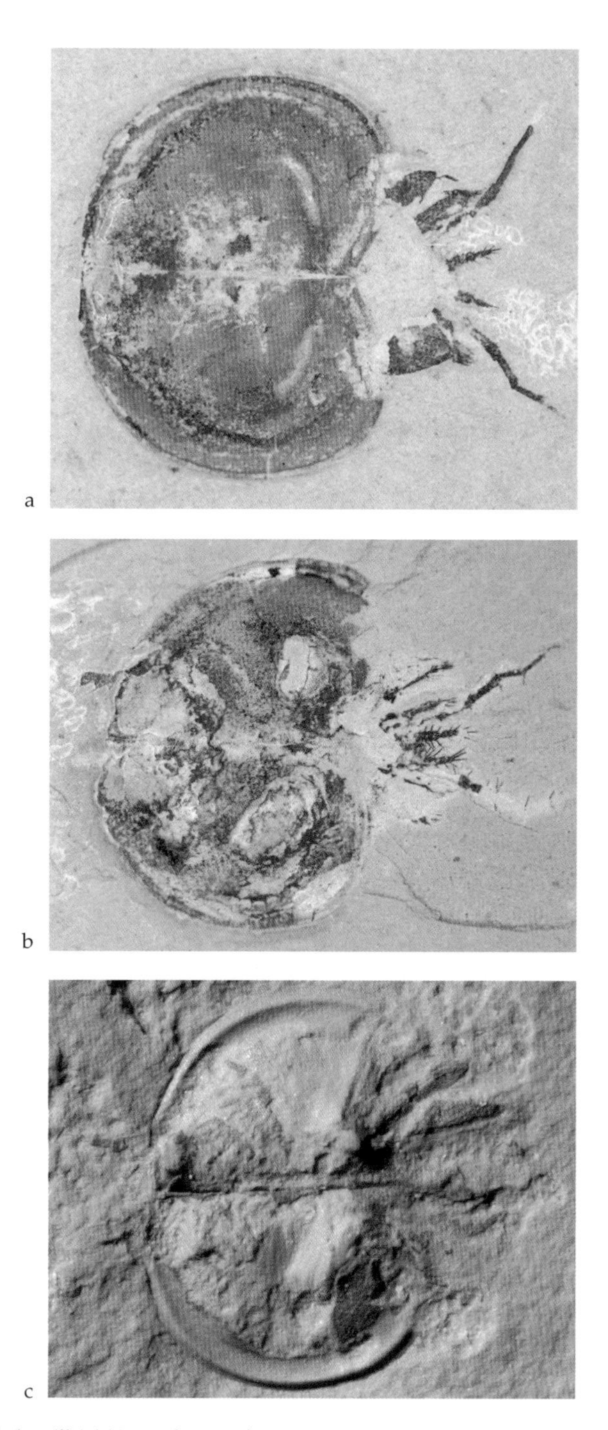

Figure 16.13 *Kunmingella douvillei*. (a) Dorsal view of open carapace (RCCBYU 10258), ×9.2; Ercaicun. (b) Dorsal view of open carapace (RCCBYU 10259), ×9.4; Ercaicun. (c) Dorsal view of open carapace (NIGPAS 78184), ×12.7; Xiaolantian.

Genus *Isoxys* Walcott, 1890

Isoxys auritus (Jiang, 1982)

I. auritus is a very common element of the Chengjiang biota, but specimens with soft parts preserved are rare. The thin, elongate and bivalved carapace can be more than 45 mm long and is extended into spines of subequal length at the anterodorsal and posterodorsal corners. An ornament of fine lines and fine reticulation occurs on the carapace.

Based largely on Chinese material, it has been demonstrated that *Isoxys* had a long segmented body, with a pair of forwardly projecting stalked eyes and short antennae, followed by a series of 14 essentially similar biramous appendages (Vannier & Chen 2000). Details of the biramous appendages are difficult to discern, but they appear to consist of short segmented endopods and longer paddle-like exopods fringed with setae. In lateral view it seems that the eyes, antennae and the setiferous tips of the exopods are the soft parts that are clearly seen to protrude outside the widely gaping bivalved carapace. This picture amends a previous reconstruction of *Isoxys* (Shu *et al.* 1995b) that was based on limited soft-part material.

The general carapace design and some aspects of appendage morphology of *Isoxys* recall the bradoriid *Kunmingella*, a group to which *Isoxys* has previously sometimes been allied simply because it has a small bivalved carapace. Differences include the internal position of the eyes and far fewer segments in *Kunmingella*. As with *Kunmingella*, the essentially similar morphology of its post-antennular limbs hints that *Isoxys* is not a very specialized or derived arthropod. Additional knowledge of the appendages, particularly the proximal parts, is needed in order to resolve the systematic position of *Isoxys*.

Paleogeographically, *Isoxys* seems to be restricted to tropical and subtropical regions, indicating possible temperature controls on its distribution (Williams *et al.* 1996). *Isoxys* was probably an active swimmer, perhaps high in the water column, thereby attesting to the early Cambrian colonization of such niches (Vannier & Chen 2000).

I. auritus is known only from the Chengjiang biota.

Key References Jiang *in* Luo *et al.* 1982, Hou 1987c, Shu *et al.* 1995b, Williams *et al.* 1996, Chen & Zhou 1997, Hou *et al.* 1999, Vannier & Chen 2000.

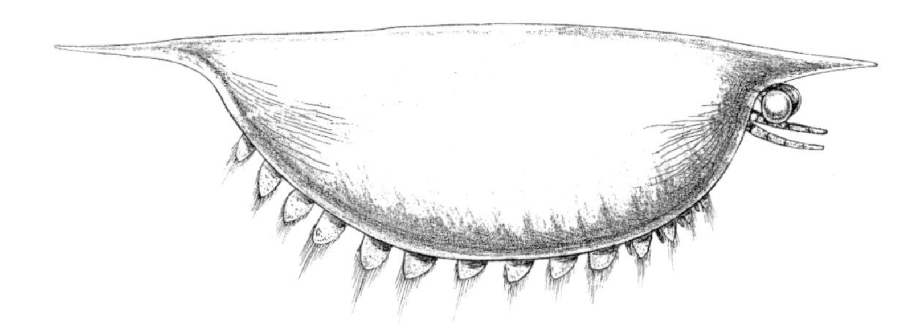

Figure 16.14 Reconstruction of *Isoxys* (based on data of Vannier & Chen 2000 and material herein).

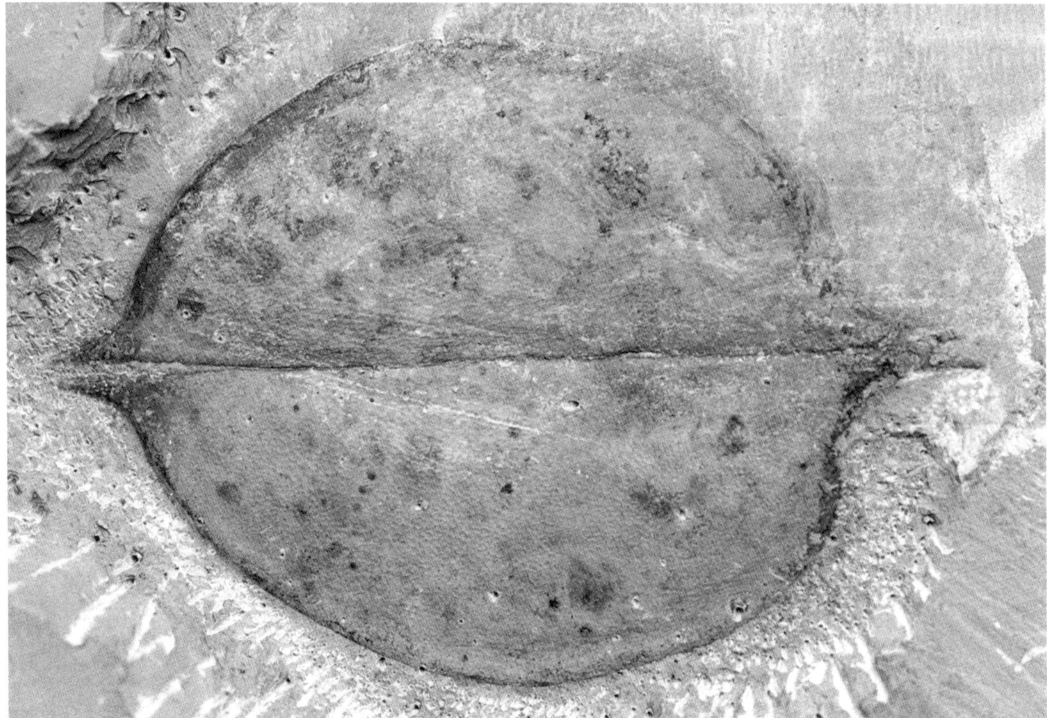

Figure 16.15 *Isoxys auritus*. (a) Lateral view (RCCBYU 10261), ×3.8; Mafang. (b) Dorsal view of open carapace (RCCBYU 10262), ×2.6; Mafang.

Isoxys paradoxus Hou, 1987

This is one of three *Isoxys* species recorded from the Chengjiang biota. It is relatively rare and, unlike its Chengjiang associates *Isoxys auritus* and *Isoxys curvirostratus*, is known only from carapaces.

The thin, elongate, bivalved carapace has a straight spine at both the anterodorsal and posterodorsal corners. The posterior spine is longer than the bivalved part of the carapace. Including both spines, carapace length can exceed 100 mm. *I. paradoxus* can easily be distinguished from *I. auritus* by the unequal and total length of its spines. *I. curvirostratus* is distinguished from other species of the genus by having a curved anterior spine (Vannier & Chen 2000).

Isoxys is a component of the earliest arthropod faunas worldwide. The genus is known from the Lower Cambrian of Spain, Siberia, South Australia and Southwest China and also from the Lower to Middle Cambrian of Laurentian North America. The ecology of *Isoxys* is discussed under *I. auritus*.

I. paradoxus is unknown outside the Chengjiang biota.

Key References Hou 1987c, Williams *et al.* 1996, Chen & Zhou 1997, Hou *et al.* 1999, Vannier & Chen 2000.

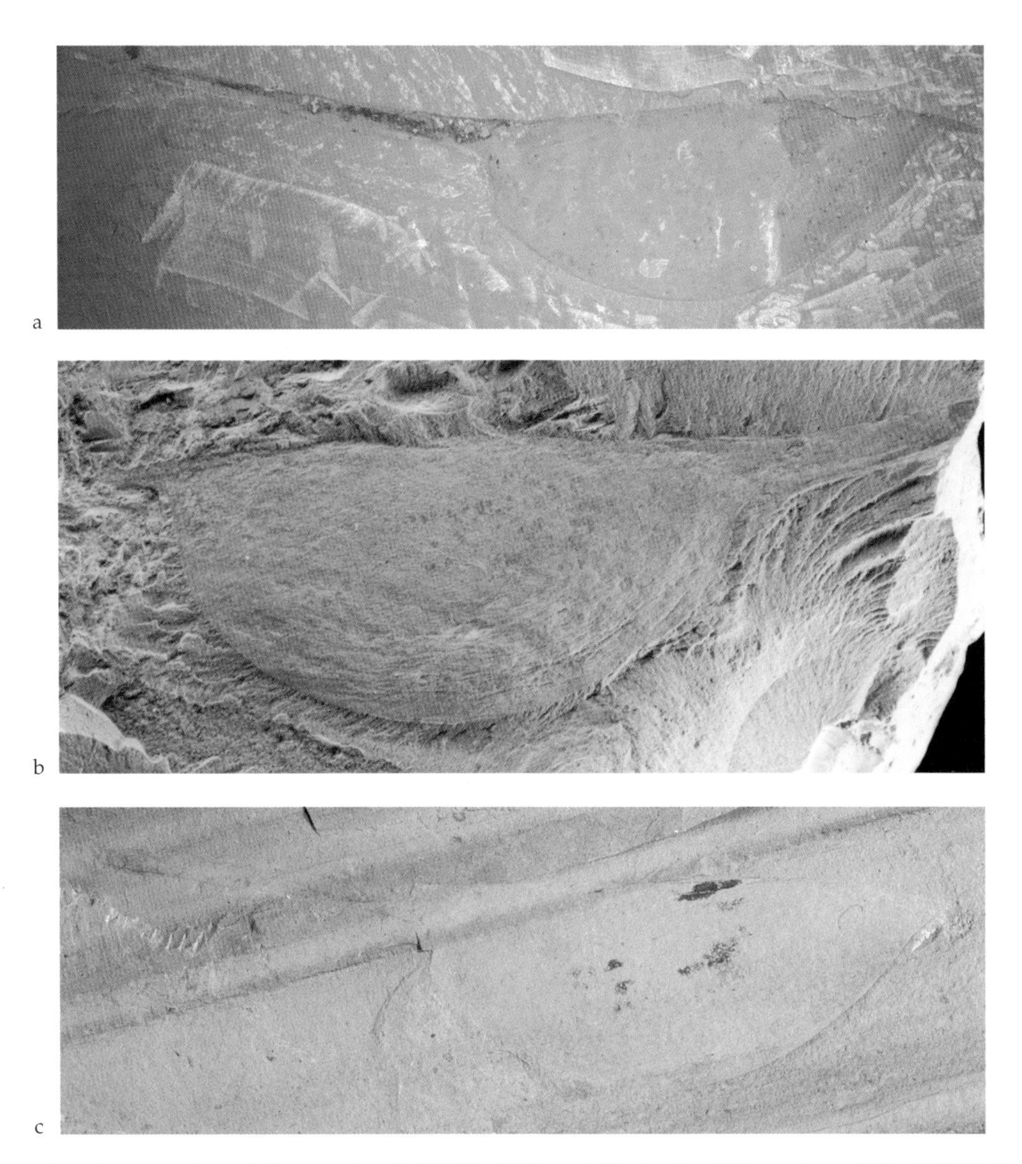

Figure 16.16 *Isoxys paradoxus*. (a) Lateral view (RCCBYU 10263), ×2.5; Xiaolantian. (b) Lateral view (NIGPAS 110829), ×1.3; Maotianshan. (c) Lateral view (RCCBYU 10313), ×2.0; Maotianshan.

Genus *Yunnanocaris* Hou, 1999

Yunnanocaris megista Hou, 1999

This arthropod is known from only two bivalved carapaces, both of which show modest relief.

Each valve is large and suboval, rounded both anteriorly and posteriorly, and has a straight and relatively short dorsal margin measuring about half the length of the valve. The paratype carapace is 71 mm long and 53 mm high. In lateral view the valves show a marked backward swing. The valve surface lacks ornament but has concentric wrinkles and irregular ridges and depressions resulting from compaction of the convex valves. No appendages or other soft parts of *Y. megista* are known.

The carapace of *Y. megista* is somewhat similar in shape to that of *Canadaspis* but is deeper posteriorly and more vaulted and it has a shorter hinge. *Y. megista* also shows some resemblance to the Chengjiang species *Waptia ovata*, from which it differs in being much larger and in having a straight rather than arched dorsal margin.

In the absence of knowledge of its soft parts, the affinities and mode of life of *Y. megista* are unclear. *Yunnanocaris* is known from only a single, Chengjiang biota species.

Key References Hou 1999, Hou *et al.* 1999.

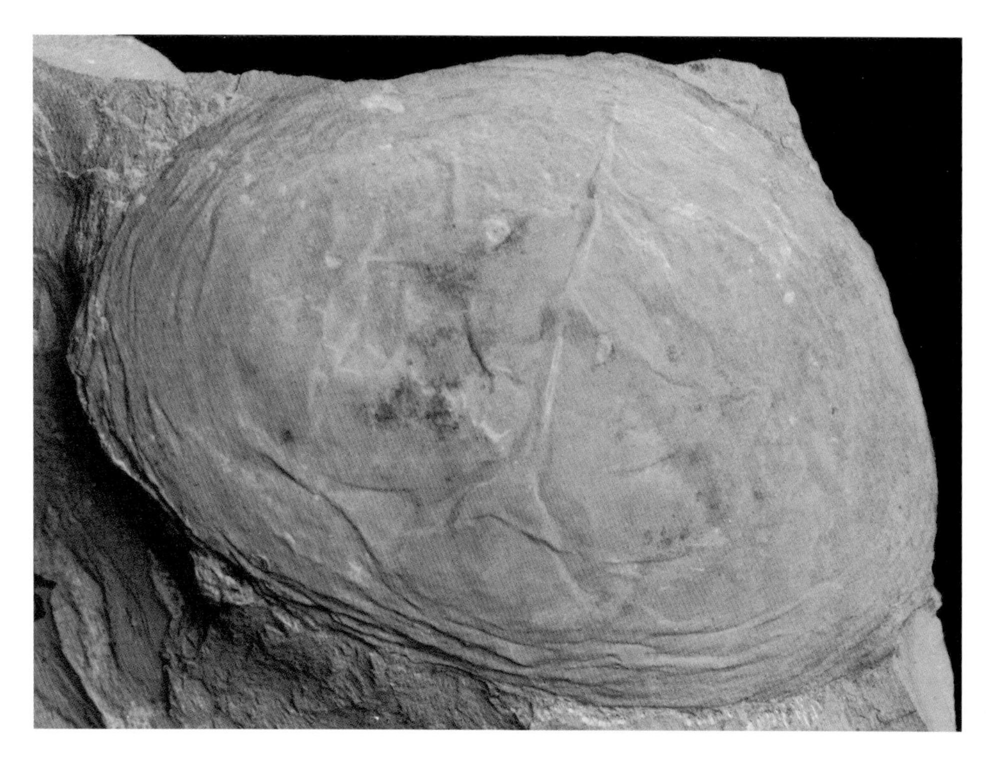

Figure 16.17 *Yunnanocaris megista*. Lateral view (NIGPAS 115415), ×1.7; Maotianshan.

Genus *Leanchoilia* Walcott, 1912

Leanchoilia illecebrosa (Hou, 1987)

This species, known from a number of localities, is one of the most common arthropods in the Chengjiang fauna. Many specimens lie on their side, others are dorsoventrally or obliquely compressed. The gut is often seen in bas-relief.

There are conflicting assessments of some of the morphological features of *L. illecebrosa* (Hou & Bergström 1997; compare Hou 1987, Chen & Zhou 1997). The head shield is succeeded by 11 segmental tergites and a flat, elongate, dagger-like telson bearing slender marginal spines. A pair of dark spots, near the front end of the head shield, probably represent the eyes. The distinctive anterior appendage has three flagella, each longer than the entire body. There are probably three additional pairs of biramous appendages in the head and one pair of biramous appendages per body segment. The endopod consists of about six simple cylindrical podomeres. The exopod has at least two parts, the distal one being paddle-like in shape with long, needle-like marginal setae.

L. illecebrosa was originally assigned to the genus *Alalcomenaeus* Simonetta, 1970, which like the morphologically similar *Leanchoilia* was first described from the Burgess Shale. It has been suggested that both genera may be represented in the Chengjiang material (see Della Cave & Simonetta 1991, Briggs & Collins 1999). *Leanchoilia* has been assigned to its own order (Hou & Bergström 1997). Cladistic analyses resolve *Leanchoilia* as an arachnomorph (Wills *et al.* 1998) or as a stem-group euarthropod close to *Alalcomenaeus* (Budd 2002). Material that may represent *L. illecebrosa* has been described from the Chengjiang biota under the names *Leanchoilia asiatica*, *Dianchia mirabilis*, *Yohoia sinensis* and possibly *Zhongxinia speciosa* and *Apiocephalus elegans* (Luo & Hu *in* Luo *et al.*, 1997, Luo & Hu *in* Luo *et al.*, 1999).

It seems likely that the flagella of the large appendages in *L. illecebrosa* were used for sensory purposes. The construction of the outer branch of the appendages indicates that the animal was probably a good swimmer. Some specimens of *L. illecebrosa* apparently have a mud-filled gut, inferring possible deposit feeding activities (Hou & Bergström 1997, Bergström 2001). For the Burgess Shale species *Leanchoilia superlata* Walcott, 1912, which is considered to have a predatorial/scavenging mode of life, such fill has been interpreted as the permineralized glands of the mid-gut, subsequently replaced by clay minerals (Butterfield 2002).

This species is known only from the Chengjiang biota.

Key References Hou 1987a, Hou & Bergström 1997, Chen & Zhou 1997, Hou *et al.* 1999.

Figure 16.18 *Leanchoilia illecebrosa*.(a) Lateral view (NIGPAS 115367), ×3.7; Jianbaobaoshan, near Dapotou. (b) Lateral view (NIGPAS 115363), ×4.2; Xiaolantian. (c) Lateral view (RCCBYU 10264), ×3.5; Mafang. (d) A cluster of specimens in lateral view (RCCBYU 10265), ×1.6; Mafang.

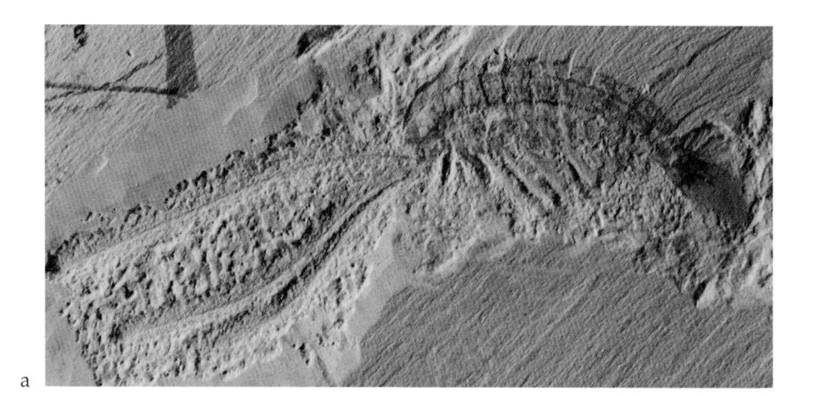
a

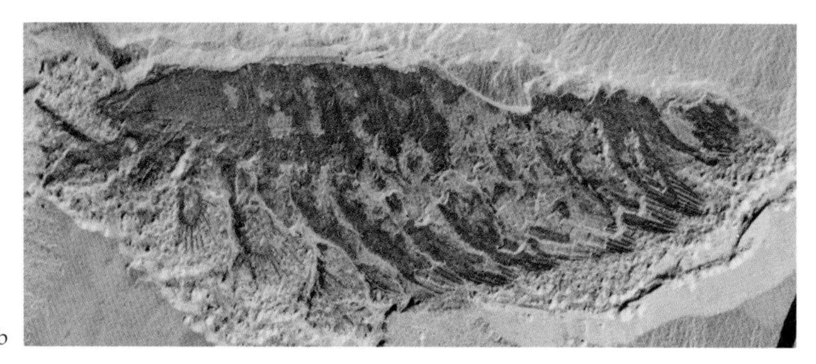
b

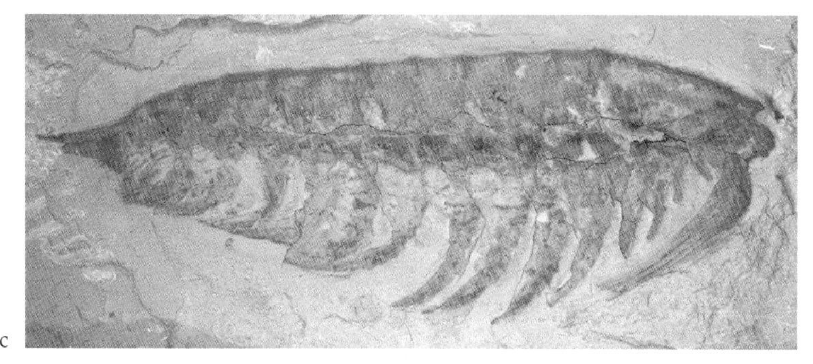
c

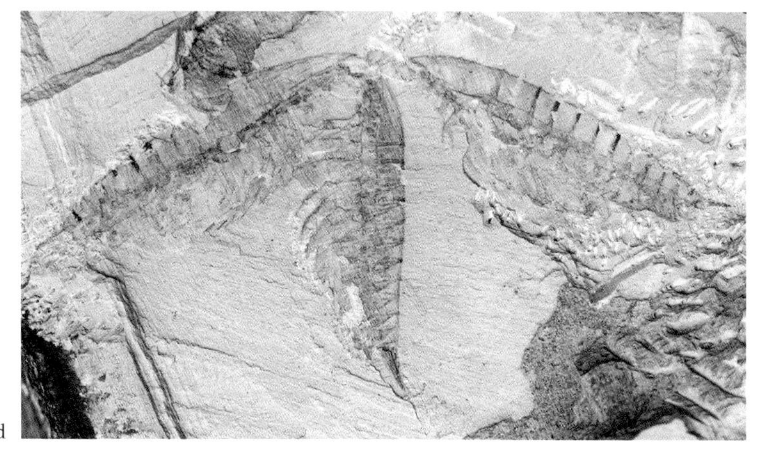
d

Genus *Jianfengia* Hou, 1987

Jianfengia multisegmentalis Hou, 1987

This species is known from about ten specimens, preserved flattened in both lateral and dorsoventral aspects. The type specimen is about 17 mm long.

J. multisegmentalis has a short head, a trunk with about 22 segmental tergites and a telson. There is a pair of eyes, which may have been on stalks just in front of the anterior margin of the head shield. A few wrinkles indicate the presence of at least three segments in the head. The head shield bends down anteriorly, is transversely vaulted, and extends into a short, probably ventrolaterally directed pleural fold. The trunk tergites are of simple shape and transversely rounded. The telson seems to be a narrow, more or less parallel-sided rod-like structure with a pointed end.

The first appendage is of "great appendage" type. It emerges from the anterior part of the head; proximally it is long and slender, distally it widens and possibly ends in an array of small spines. The position and possibly unbranched character of this appendage may indicate that it represents a first antenna; it may not correspond to the "great appendage" of other species. The head and the trunk contain a total of 25 other pairs of (biramous) appendages. The endopods are long and slender, consisting of many podomeres and a spine-like tip distally. Each exopod appears to be an elongate flap that carries long setae on the outer edge. The alimentary canal is often preserved as a darker tract in the trunk region and as an oval area in the head.

In having anterior "great appendages" this species seems to be related to *Leanchoilia* and allied taxa. It appears to be closely comparable to the Burgess Shale *Yohoia tenuis* Walcott, 1912, which notably differs in having extremely large eyes and much fewer trunk segments, the last three of which have lost their appendages. A cladistic analysis has resolved *Jianfengia* as an upper stem-group euarthropod closely related to *Yohoia* (Budd 2002).

The ecology of *J. multisegmentalis* is poorly understood. The pointed endopods seem to be unsuitable for walking. The exopods may have aided locomotion. The "great appendage" may have been used in grasping prey.

This animal is unknown outside the Chengjiang biota.

Key References Hou 1987a, Chen & Zhou 1997, Hou *et al.* 1999.

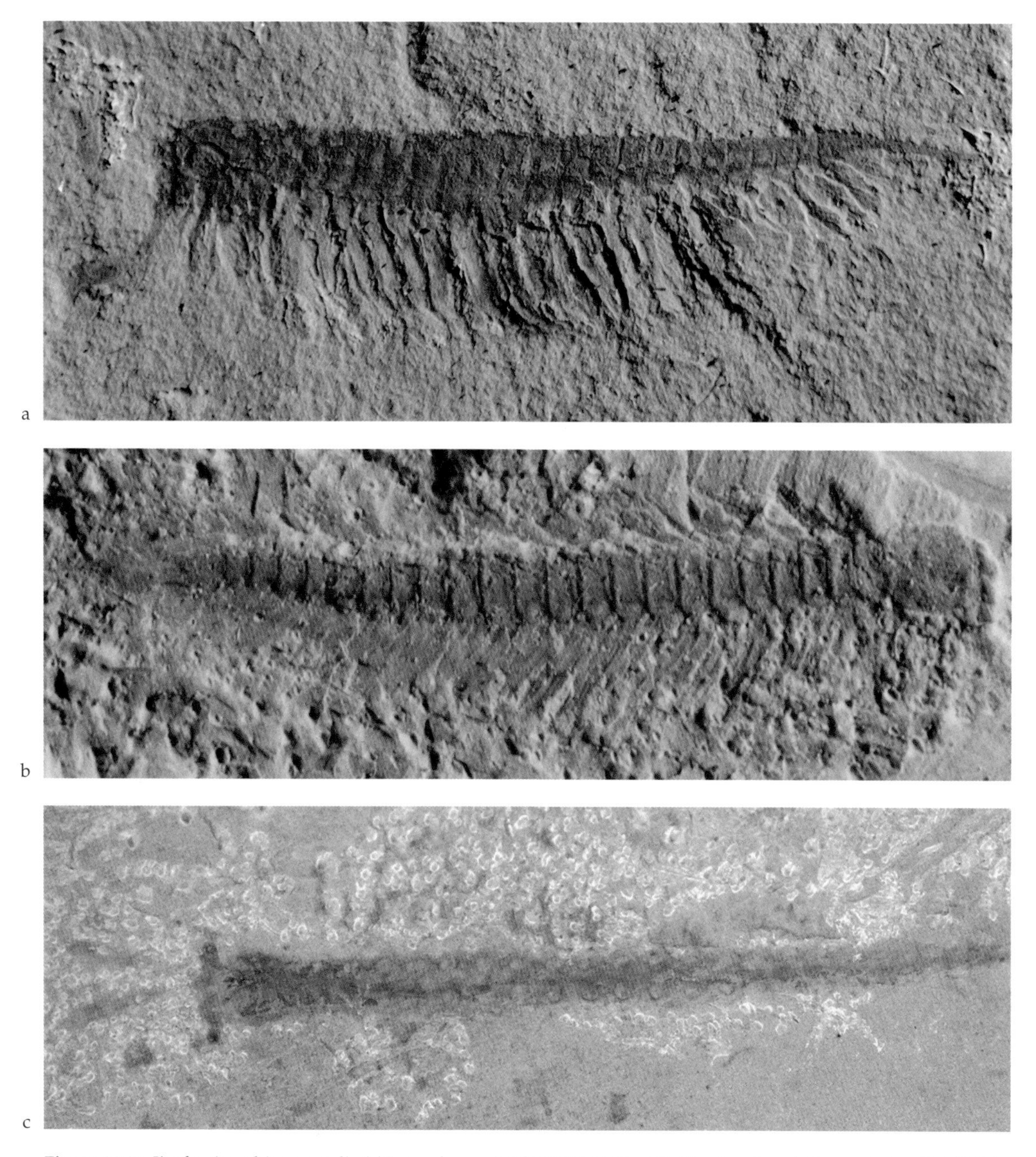

Figure 16.19 *Jianfengia multisegmentalis*. (a) Lateral view (NIGPAS 100123), ×7.0; Maotianshan. (b) Lateral view (RCCBYU 10266), ×7.0; Maotianshan. (c) Dorsal view (RCCBYU 10267), ×7.1; Maotianshan.

Genus *Tanglangia* Luo & Hu *in* Luo, Hu, Chen, Zhang & Tao, 1999

Tanglangia caudata Luo & Hu *in* Luo, Hu, Chen, Zhang & Tao, 1999

T. caudata is relatively uncommon. Specimens are flattened, up to about 35 mm long, and in overall morphology resemble the Chengjiang species *Jianfengia multisegmentalis*.

The narrow, elongate body is divided into a head shield, trunk and a telson. The head has a subelliptical outline and bears a frontal pair of "great appendages", behind which there are possibly three pairs of biramous appendages. The "great appendages" are robust, project in front of the head shield and appear to terminate in a cluster of small spines. The trunk, which bears pairs of biramous appendages, is divided into about 13 segmental tergites. The simple, narrow tail is almost as long as the entire trunk region. The material herein shows a pair of eyes projecting in front of the head shield.

It seems that *T. caudata* is related to *J. multisegmentalis* and allied forms bearing "great appendages" in the frontal part of the head. It differs from the latter species by its much larger overall size, and by having much fewer trunk segments and a longer telson.

The ecology of *T. caudata* is poorly understood. The species is known from localities in the Haikou and Chengjiang areas.

Key References Luo & Hu *in* Luo *et al.* 1999.

Figure 16.20 *Tanglangia caudata*. (a) Dorsolateral view (RCCBYU 10268), × 4.1; Ma'anshan. (b) Detail of the head and part of the trunk (RCCBYU 10268), × 17.5. (c) Lateral view (RCCBYU 10269), × 8.4; Mafang.

a

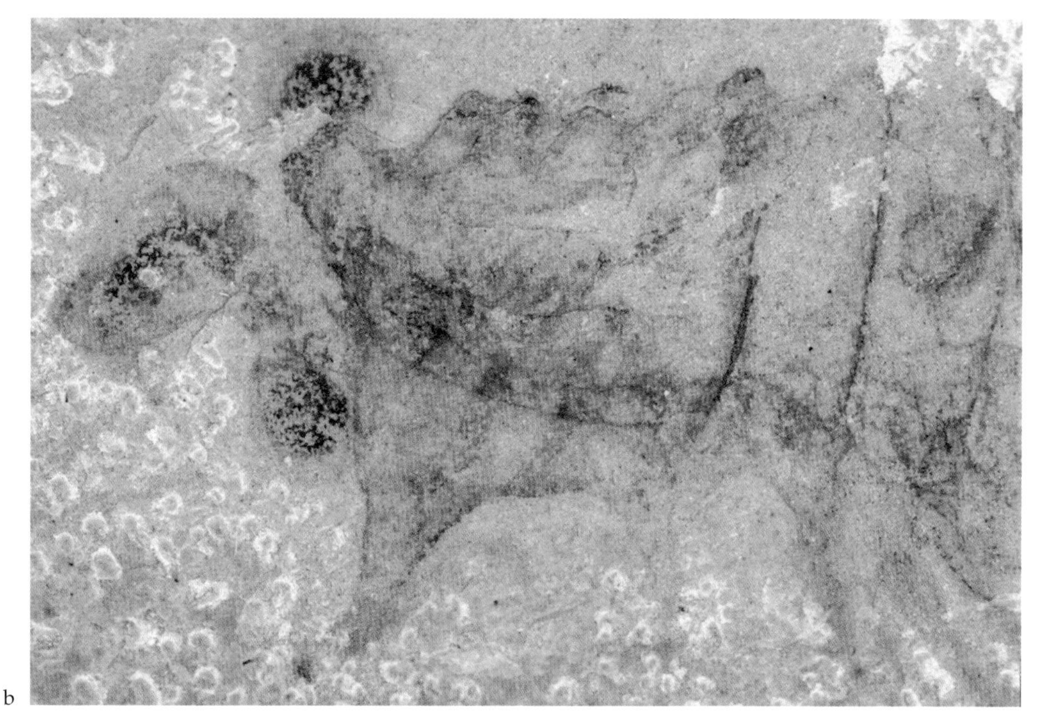

b

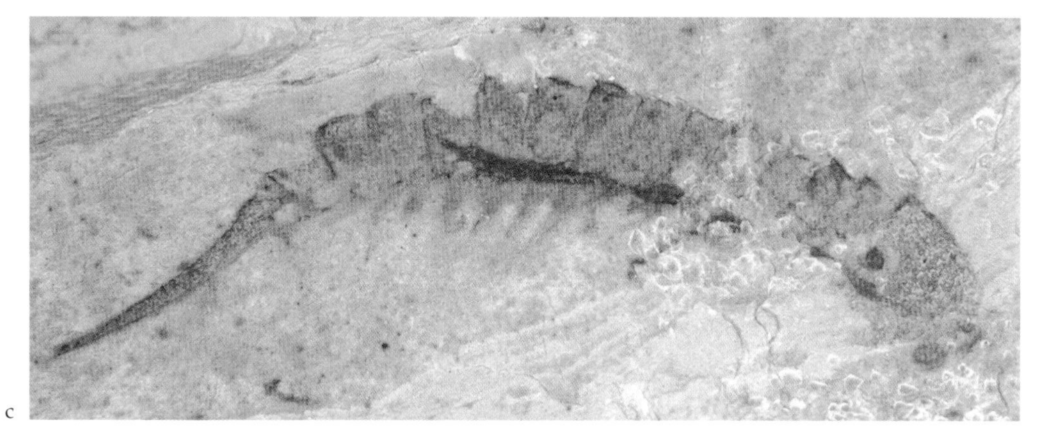

c

Genus *Fortiforceps* Hou & Bergström, 1997

Fortiforceps foliosa Hou & Bergström, 1997

This is a rare species, known from fewer than ten specimens. The exoskeleton and appendages are preserved compressed laterally, dorsoventrally or obliquely.

The body is long and slender, reaching a length of about 4 cm excluding appendages. Large stalked eyes emerge from under the front edge of the short head shield. The trunk has 20 segmental tergites with short pleural spines and there is a characteristic broad, fan-like telson. Possible antennae occur in some specimens and, if correctly interpreted, are slender and have a club-like termination. The possibly second pair of appendages, the "great appendages", is robust and each displays single spines on the inner side of each of the three distal podomeres. Three additional pairs of biramous appendages occur on the head and one pair is carried on each trunk segment. The inner, "walking" branch of these appendages consists of about 15 short podomeres and tapers to almost a point. The paddle-shaped exopod has a pattern similar to that in *Canadaspis*: a proximal part is surrounded by a distal field with faint radial lines and a dense array of marginal bristles. The tail fan, consisting of a tripartite median region flanked by a pair of longer lateral blades, has fine bristles along its posterior edge.

This species appears to belong to the *Leanchoilia* group of arthropods, characterized by the development of a strong pair of frontal limbs, the "great appendages". Hou & Bergström (1997) established an order and family based on *Fortiforceps*. A recent analysis, based particularly on head characteristics, documents *Fortiforceps* as one of several closely related Cambrian stem-group euarthropods (Budd 2002).

The spiny and well-developed second appendages no doubt were adapted to grasp food, and the animal was probably carnivorous. The limb morphology suggests that it was likely to be a swimmer. A simple straight intestine can be traced from the middle of the head to the end of the trunk, but does not show any filling.

The species is known only from localities in the Chengjiang area.

Key References Hou & Bergström 1997, Hou *et al.* 1999, Budd 2002.

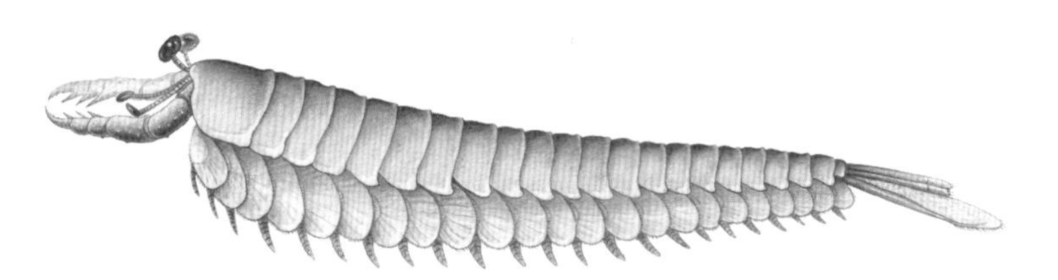

Figure 16.21 Reconstruction of *Fortiforceps foliosa* (after Hou & Bergström 1997).

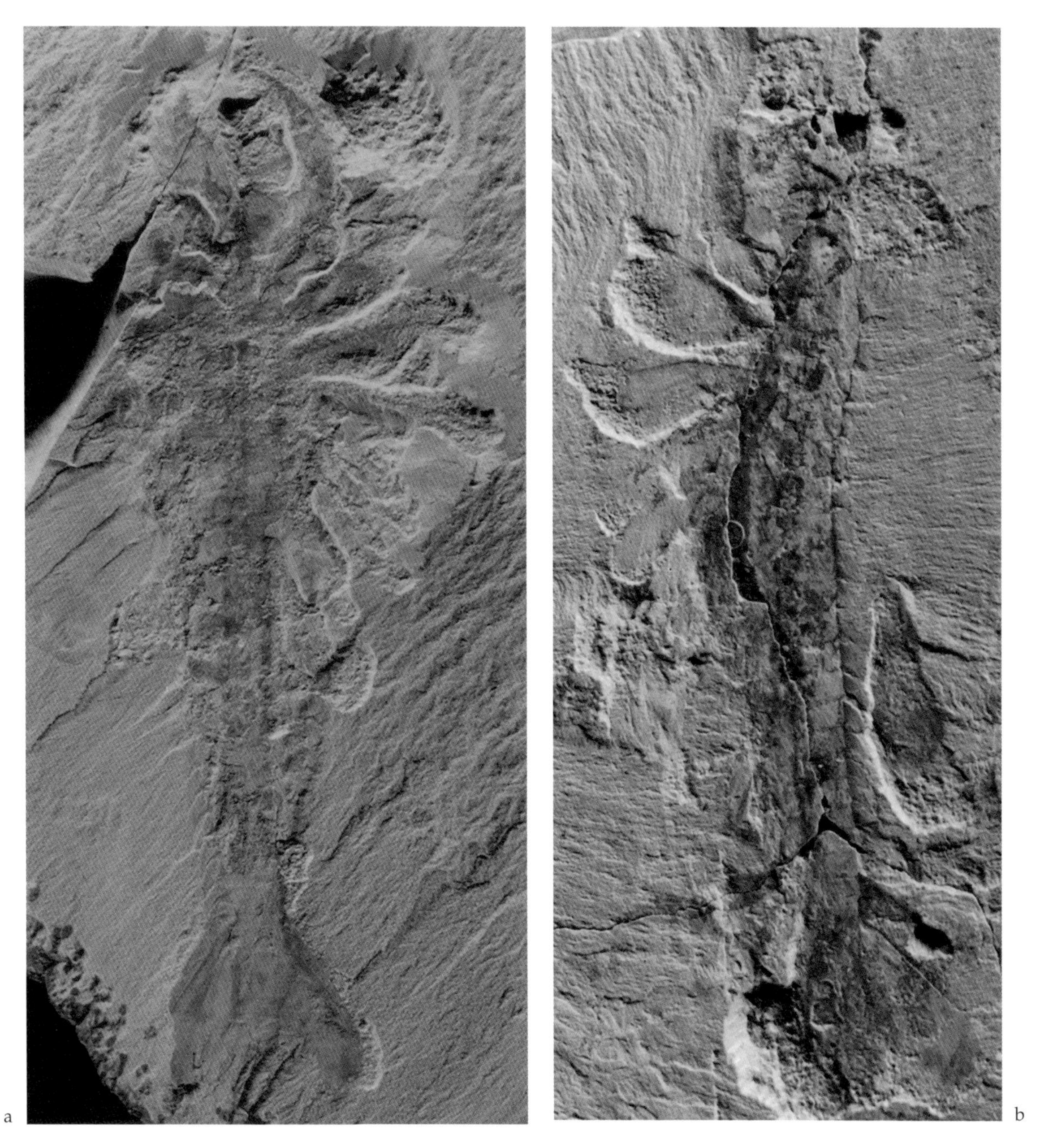

Figure 16.22 *Fortiforceps foliosa*. (a) Ventral view (NIGPAS 115373), ×3.8; Maotianshan. (b) Lateral view (NIGPAS 115372), ×3.8; Maotianshan.

Genus *Occacaris* Hou, 1999

Occacaris oviformis Hou, 1999

This species is known from a single specimen. Its carapace seems to be smooth, thin and unmineralized.

The carapace is bivalved, about 8mm long and 6mm high and, as alluded to in the species name, is approximately egg-shaped; it covers the head and anterior parts of the body. Two or three trunk tergites and a partially preserved telson extend behind the carapace. The first of the trunk tergites may be articulated directly to the carapace. Pairs of possible first antennae, consisting of at least 15 annuli, and the second antennae ("great appendages") protrude beyond the anterior margin of the valves. As in the Chengjiang associate *Forfexicaris valida* the robust "great appendage" has spines distally, but in *Occacaris* the spines are paired. Above and just outside the carapace are two globular structures, which are most likely a pair of stalked eyes. Behind these, projecting outside the anterior part of the carapace, somewhat shorter multisegmented limbs presumably represent the inner branch of head and trunk appendages. More posteriorly, the remains of three setiferous flaps are considered to be parts of the exopods of appendages.

O. oviformis was erected as the sole member of the Family Occacarididae. The "great appendage" and at least the endopod of the other appendages are closely comparable with corresponding structures in *F. foliosa* and hint that these taxa are related. Analysis based primarily on features of the head has placed *Occacaris* and *Fortiforceps* as closely related taxa in a group of Cambrian stem-group euarthropods (Budd 2002).

The spinose "great appendages" may have been used for grasping small prey, but there is no other evidence to help to interpret the feeding mode of this species. As knowledge of most of the appendages is scant, its mode of locomotion is also uncertain. The only known specimen is from the Chengjiang area.

Key References Hou 1999, Hou *et al.* 1999.

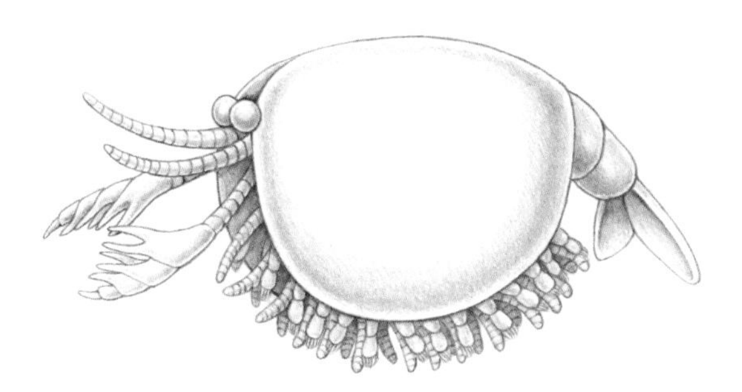

Figure 16.23 Reconstruction of *Occacaris oviformis*.

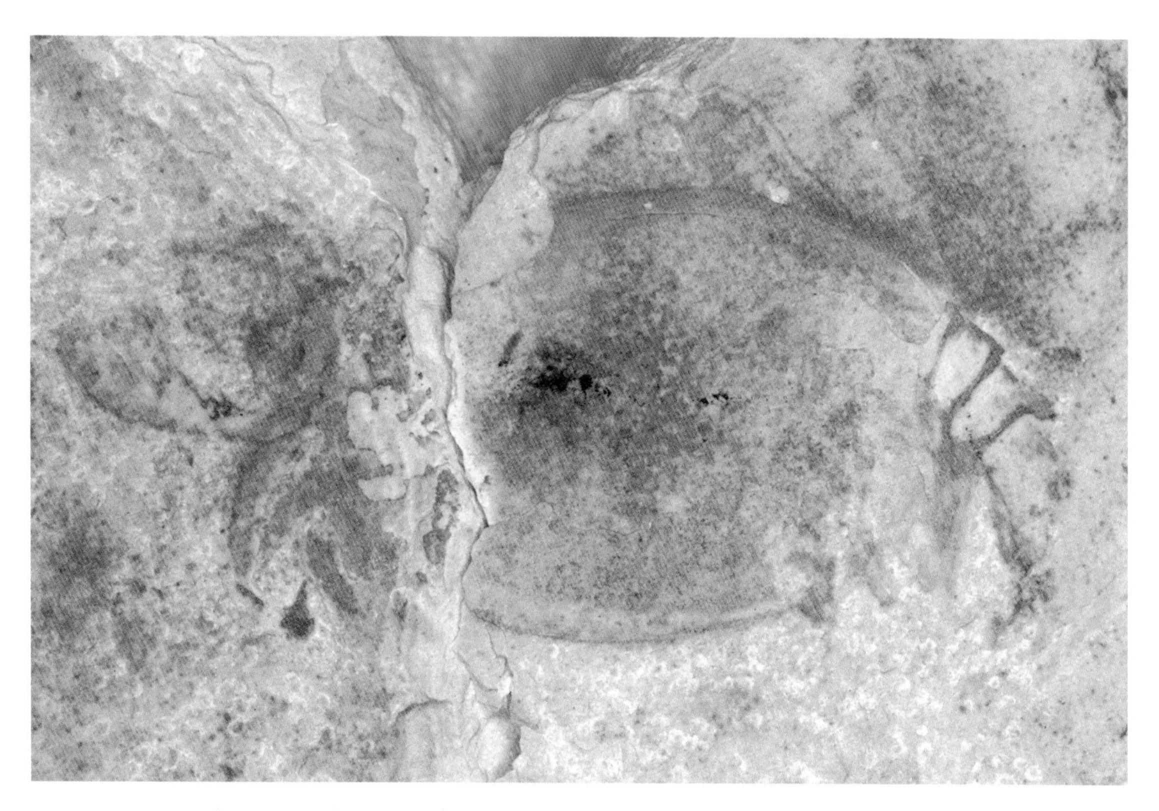

Figure 16.24 *Occacaris oviformis*. Lateral view (NIGPAS 115408), ×11.9; Maotianshan.

Genus *Forfexicaris* Hou, 1999

Forfexicaris valida Hou, 1999

Only two specimens of this species are known. They are laterally flattened, with just a few soft parts preserved in front of the thin, unmineralized valves.

The carapace of the type specimen is about 1.5 cm long and 1.3 cm high, subcircular in outline with a straight dorsal margin. Anteriorly and outside the valves there is a pair of stalked eyes. The strongly developed "great appendages" appear to be much like those in *Occacaris* from Chengjiang, except that there are single rather than paired spines on the inner side of three distal podomeres. Fragmentary remains of paddle-shaped elements in front of the valves, each bearing a row of well-developed marginal setae, are presumed to be part of the outer branch of appendages. The corresponding endopods are unknown, as is the morphology of the hind part of the body.

F. valida was established as the only member of the Family Forfexicarididae. The species can be tentatively placed among other Chengjiang forms with large frontal appendages, for instance *Fortiforceps foliosa* and *Occacaris oviformis*. *F. valida* clearly differs from *O. oviformis* in having much deeper valves and, notwithstanding factors of preservation, in seemingly lacking a "first antenna" and a trunk region outside the valves.

With the limited material and incomplete knowledge of the appendages, little can be deduced concerning the ecology of *F. valida*. It seems likely, however, that the robust "great appendages" could have been used for grasping purposes. The species is known only from the Chengjiang area.

Key References Hou 1999, Hou *et al.* 1999.

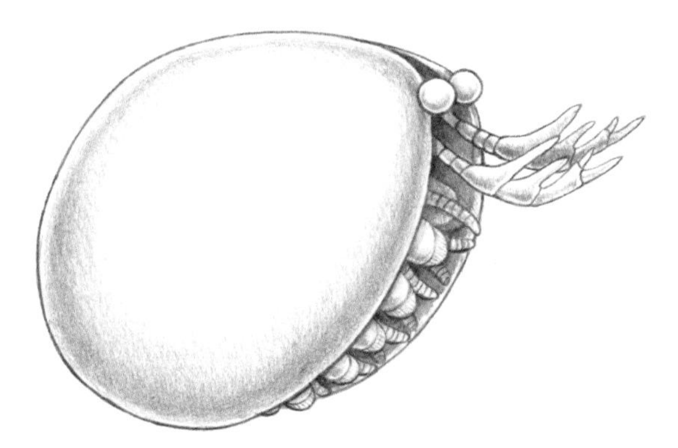

Figure 16.25 Reconstruction of *Forfexicaris valida* (the form of the endopods is hypothetical).

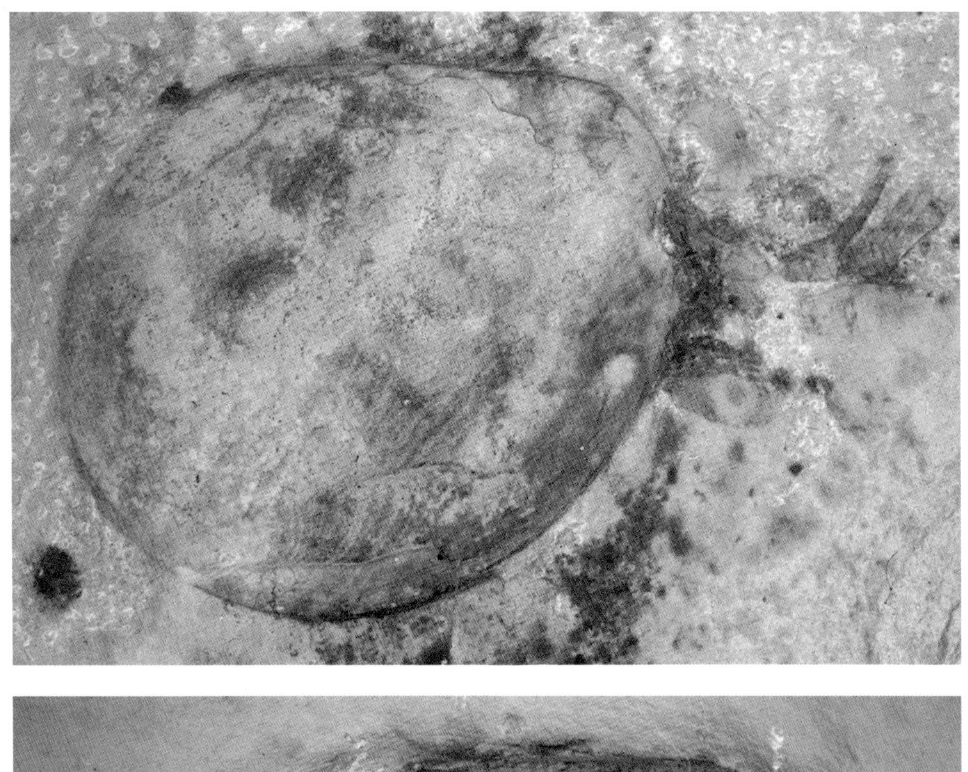

a

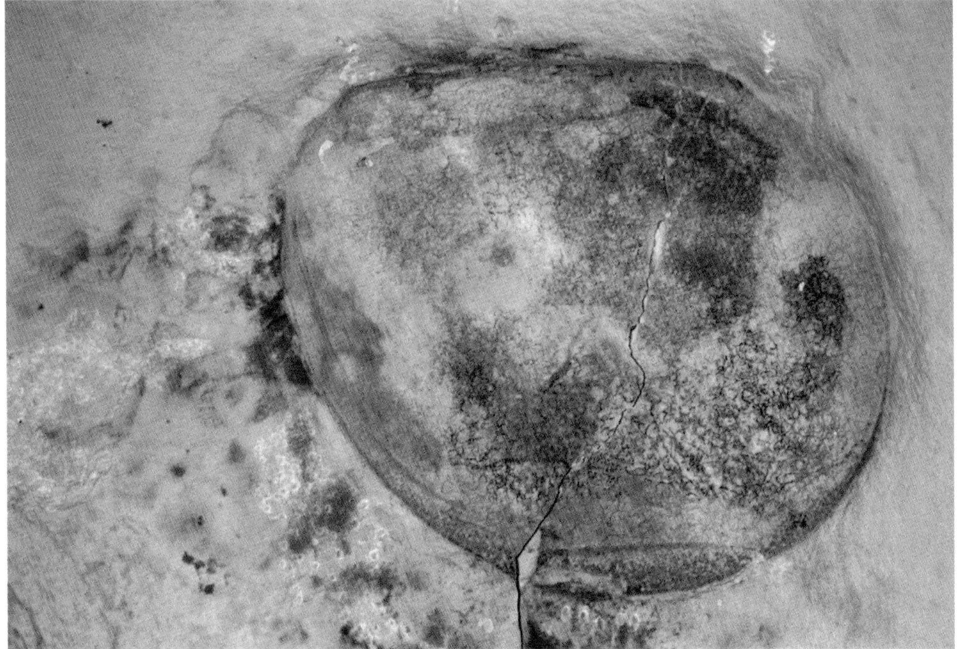

b

Figure 16.26 *Forfexicaris valida*. (a) Lateral view, part (NIGPAS 115409a), × 11.8; Xiaolantian. (b) Lateral view, counterpart (NIGPAS 115409b), × 11.8.

Genus *Pseudoiulia* Hou & Bergström, 1998

Pseudoiulia cambriensis Hou & Bergström, 1998

Only two specimens of *P. cambriensis* are known. Both are more or less laterally compressed and well preserved but incomplete. The head is unknown.

This species has an elongate, myriapod-like body. The type specimen, which is about 37 mm long and 5 mm high, shows 31 tergites, plus what appears to be the terminal element that may be incomplete. The body seems to have been strongly vaulted in cross-section. The appendages are poorly preserved. Setiferous fragments and elongate flaps bearing marginal setae represent parts of the exopods. The inner branches of the limbs are not seen.

Pseudoiulia seems to be quite different from all of the other arthropods in the Chengjiang biota. It does not seem to have any close relatives. The setation of the exopod makes it unlikely that it is close to the Chengjiang taxa *Fuxianhuia* and *Chengjiangocaris*. It can hardly be placed among trilobite-like or crustacean-like arthropods. It is therefore tentatively allied with the "great appendage" arthropods, some of which, such as *Jianfengia* and *Fortiforceps*, are notably long-bodied.

There is insufficient evidence to deduce the mode of life of *P. cambriensis*. The species is known only from the Chengjiang biota.

Key References Hou & Bergström 1998, Hou *et al.* 1999.

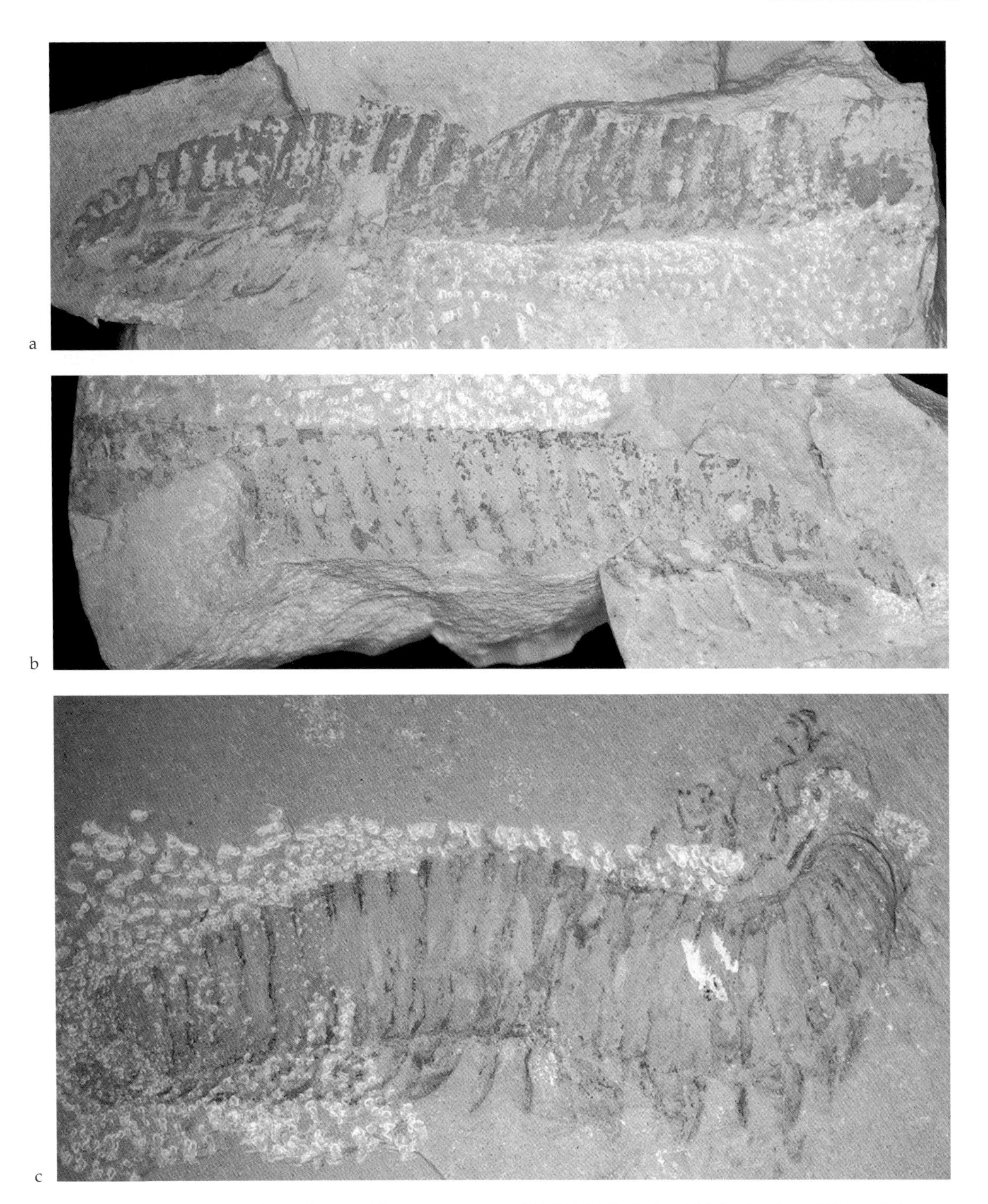

Figure 16.27 *Pseudoiulia cambriensis*. (a) Lateral view, part (NIGPAS 115420a), ×4.0; Maotianshan. (b) Lateral view, counterpart (NIGPAS 115420b), ×4.0. (c) Dorsolateral view (RCCBYU 10270), ×5.0; Mafang.

Genus *Waptia* Walcott, 1912

Waptia ovata (Lee, 1975)

This is a common species, known mostly from the wrinkled remains of its carapace. Some carapaces are found in clusters.

The carapace is folded about a median line and is typically about 1 cm long. Projecting anteriorly from below the carapace there is a pair of compound eyes and a pair of very long, multisegmented uniramous antennae. Extending behind the carapace there are six or seven limbless segments of the trunk, ending in a flattened, bilobed telson. The segments covered by the valves are reported to have pairs of biramous appendages, each limb consisting of filamentous exopods and multisegmented endopods with a terminal claw (*Waptia* of Chen & Zhou 1997).

This species was originally placed within the "ostracod" genus *Mononotella* and was later (Hou & Bergström 1991) referred to a new genus, *Chuandianella*. However, finds of body parts other than the carapace shows that it is similar to the Burgess Shale species *Waptia fieldensis* Walcott, 1912. The affinity of *Waptia* is uncertain; its assignment as a true crustacean (e.g. Wills 1998a, Wills *et al.* 1998) has been disputed on the basis of appendage morphology (Hou & Bergström 1997, Walossek 1999).

Waptia seemingly did not have appendages that were strongly developed for comminuting food. Briggs *et al.* (1994) considered that it was probably benthic, possibly feeding on organic particles in the sediment. The absence of any sediment fill in the gut region of *W. ovata*, traced as a dark band down the trunk, hints that it selected its food carefully rather than just indiscriminately ingesting sediment as was possibly the case in many of its arthropod contemporaries (Hou & Bergström 1997). *Waptia* specimens are also known from supposed coprolites found in the Chengjiang Lagerstätte (Chen & Zhou 1997).

The species has been reported from the Lower Cambrian of Yunnan, Sichuan, Ghizhou and southern Shaanxi provinces, China, but only the material from the Chengjiang biota contains soft parts.

Key References Hou & Bergström 1991, Chen & Zhou 1997, Hou & Bergström 1997, Hou *et al.* 1999.

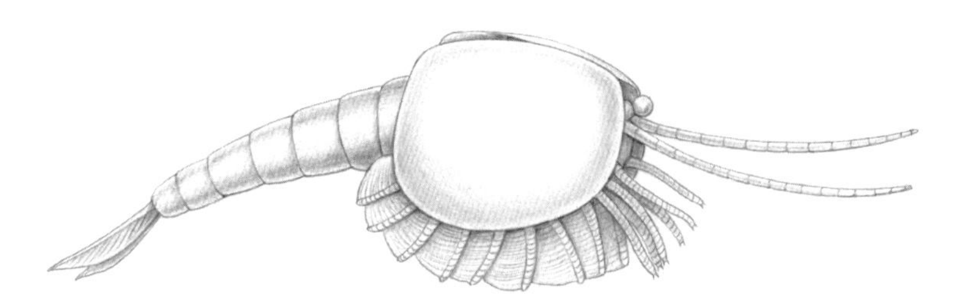

Figure 16.28 Reconstruction of *Waptia ovata* (the post-antennal appendages have not been studied in detail and their reconstruction is tentative).

Figure 16.29 *Waptia ovata*. (a) Dorsolateral view (RCCBYU 10271), ×3.5; Maotianshan. (b) Dorsolateral view (NIGPAS 115376), ×4.1; Maotianshan. (c) Dorsolateral view (RCCBYU 10272), ×5.6; Mafang.

Genus *Clypecaris* Hou, 1999

Clypecaris pteroidea Hou, 1999

Originally known from only one specimen, more material including soft parts of this species has now been found but awaits detailed study.

The thin, unmineralized bivalved carapace, about 5 mm long in the type specimen, covers the head and part of the body. As originally described, paired stalked eyes and annulate antennae project anteriorly from under the carapace. Behind the presumed head region there are about 19 segments, about eight of which are covered by the valves. All but the last three of these segments appear to bear a pair of probably biramous appendages. The details of these limbs are unclear, although the supposed exopods seem to be slender and have setae along the margin. The number of post-antennal cephalic appendages is uncertain, but considering the length of the head it is likely that there are several pairs. The telson has a pair of wing-like rami. The gut can be clearly traced, in many cases in relief, from the anterior cephalic region to the last segment of the trunk.

Clypecaris was erected as the only member of the Family Clypecarididae. The genus is comparable to the Burgess Shale genera *Waptia* and *Plenocaris* (both Walcott 1912) in general form but differs, for example, in the disposition of the appendages. The detailed morphology of the appendages of *C. pteroidea* is obscure and therefore judgement on its affinity is difficult. A recent analysis resolves *Clypecaris* as one of several closely related Cambrian stem-group euarthropods (Budd 2002). The Chengjiang arthropod *Ercaicunia multinodosa* appears to be a possible synonym of *C. pteroidea*.

The gut of *C. pteroidea* is reported to be filled with fine sediment like that surrounding the fossil and the species has been considered to be a deposit feeder (Hou 1999). Its large, flat terminal rami and many biramous appendages hint that it may also have been adapted for swimming.

The species is rare in the Chengjiang area but is more common near Haikou, Yunnan Province.

Key References Hou 1999, Hou *et al.* 1999, Budd 2002.

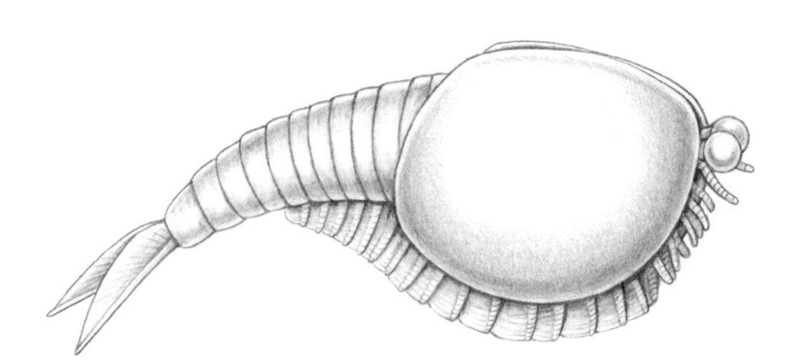

Figure 16.30 Reconstruction of *Clypecaris pteroidea*.

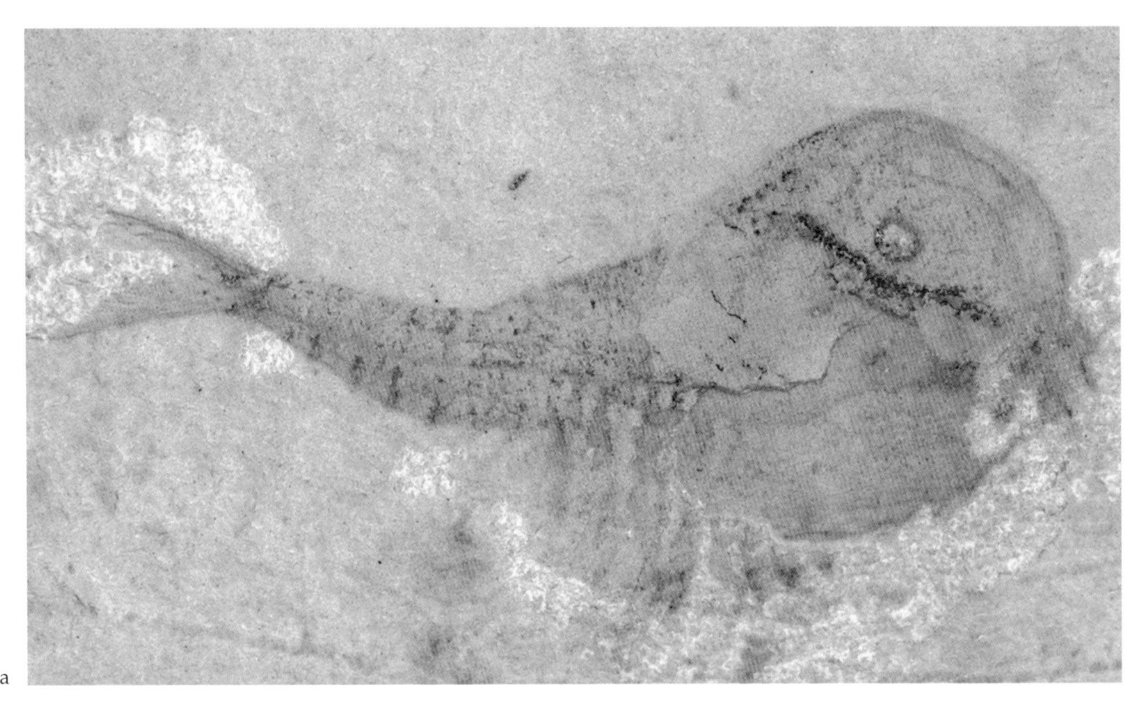

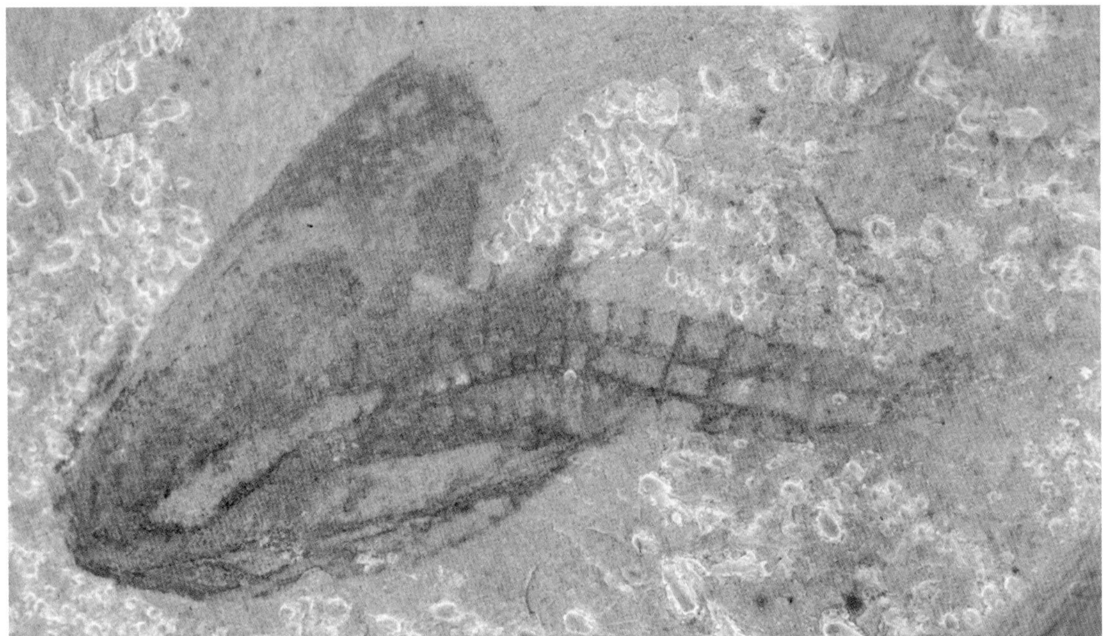

Figure 16.31 *Clypecaris pteroidea*. (a) Dorsolateral view (NIGPAS 115413), × 12.0; Xiaolantian. (b) Dorsolateral view (RCCBYU 10273), × 14.0; Mafang.

Genus *Combinivalvula* Hou, 1987

Combinivalvula chengjiangensis Hou, 1987

This species is known from several specimens, typically preserved with a little relief and in dorsal aspect. The compacted carapace is usually strongly wrinkled, indicating that in life it was probably thin and unmineralized.

The carapace is elongate, up to about 15 mm long and vaulted. It appears to be wider and deeper in the anterior part than posteriorly. A furrow marks the dorsal junction of the two sides of the carapace, but only posteriorly. This feature facilitates identification of this species even if the fossil is strongly distorted. It also implies that the carapace was not designed to open and close. Little is known of the soft parts. Dark patches mark the position of large, paired eyes in front of the carapace, and at least three fairly long and narrow trunk tergites extend out from under the rear end of the carapace.

Since *Combinivalvula* is imperfectly known, its affinities and ecology are unclear. Herein it is placed near to the waptiids because it has a vaulted carapace from which the trunk protrudes. The proportions of the carapace, however, make a close affinity with the waptiids questionable.

C. chengjiangensis is known only from the Chengjiang biota.

Key References Hou 1987c, Hou *et al.* 1999.

a

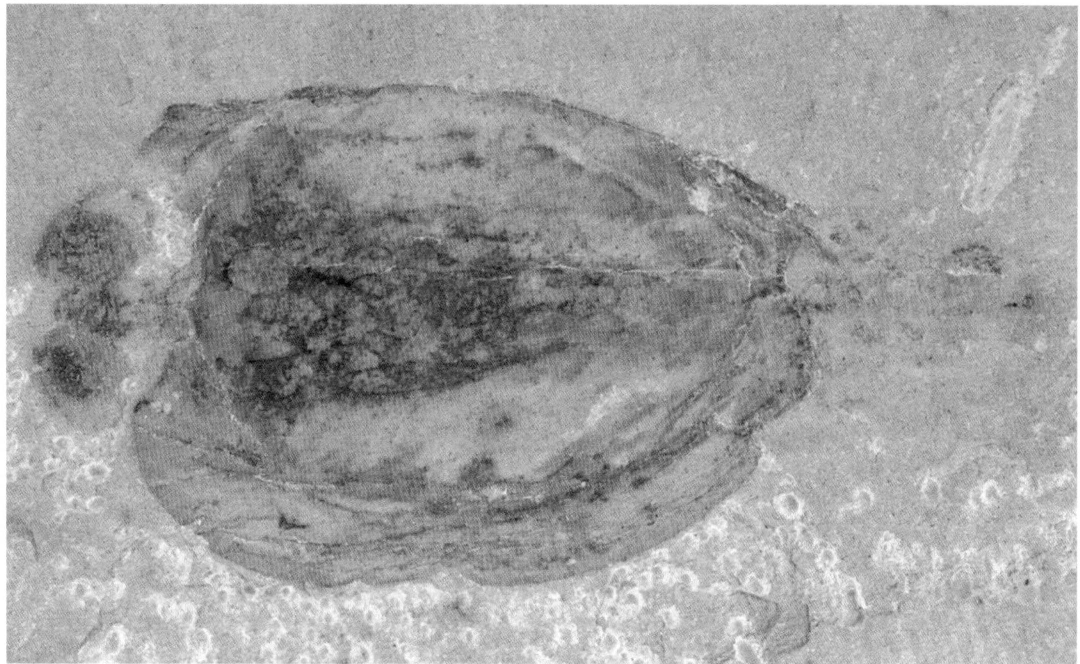

b

Figure 16.32 *Combinivalvula chengjiangensis*. (a) Dorsal view (NIGPAS 100156), × 4.9; Maotianshan. (b) Dorsal view (NIGPAS 115416), × 14.5; Xiaolantian.

Genus *Odaraia* Walcott, 1912

Odaraia? eurypetala Hou & Sun, 1988

This species is known in the literature only from the original, incomplete, flattened specimen, in which just the posterior part of the trunk is preserved. The numerous short trunk segments are succeeded by a long telson bearing a pair of broad and probably articulating blade-like rami terminally. Additional material (Hou *et al.* unpublished) allows a reconstruction of this species with paired frontal appendages and large stalked eyes projecting from a carapace more than 30 mm long. The carapace covers several tens of short segments with slender biramous appendages.

The morphology of the carapace and tail region recalls the Burgess Shale species *Odaraia alata* Walcott, 1912. *Odaraia* has a long carapace, which is also the case in the Chengjiang species *Pectocaris spatiosa*. *O.? eurypetala* and *P. spatiosa* may be more closely related than is indicated by their current assignment to separate genera. *Odaraia* has been considered to have crustacean affinities (Wills 1998a, Wills *et al.* 1998). The long terminal element and the short trunk segments are features characteristic of many branchiopod crustaceans. The poorly documented Chengjiang arthropod *Glossocaris oculatus* has a carapace similar to that of *O.? eurypetala*, and may be the same species.

The appendage morphology suggests that *O.? eurypetala* is likely to be a swimmer. The species is restricted to the Chengjiang biota.

Key Reference Hou & Sun 1988.

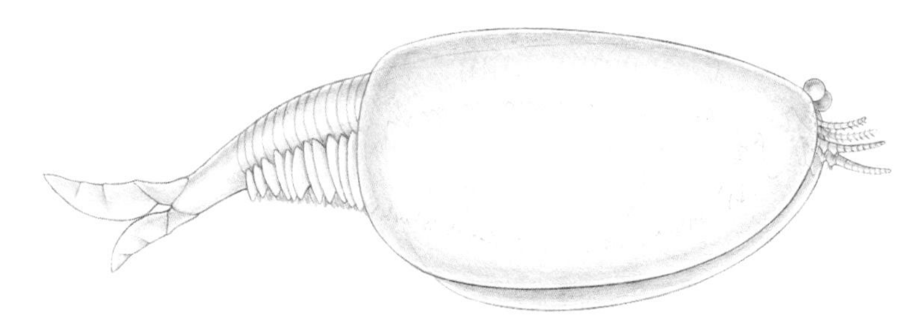

Figure 16.33 Reconstruction of *Odaraia? eurypetala*.

Figure 16.34 *Odaraia*? *eurypetala*. Lateral view (RCCBYU 10173), ×2.9; Mafang.

Genus *Pectocaris* Hou, 1999

Pectocaris spatiosa Hou, 1999

Only a few specimens of *P. spatiosa* have been found. The elongate carapace tapers gently anteriorly. The posterior end and part of the dorsal margin of the largest carapace are not preserved, but its total dimensions are estimated to be a little over 90 mm long and 50 mm high. It is one of the largest bivalved arthropods of the Chengjiang biota.

Compound eyes extend in front of the carapace, and are also found retracted. At the front of the head are the remains of the proximal parts of annulate antennae and, behind, possibly part of the gut. The carapace encloses what are interpreted to be at least 50 segments, some of which have associated biramous appendages preserved, but the details of these limbs are difficult to discern. Each endopod seems to be long and narrow and to consist of many podomeres. Several delicate setae of the exopods are also preserved. The four posteriormost segments are slightly longer than the others and appear to lack appendages. Protruding behind the carapace is part of the trunk, comprising many short segments, and a long telson ending in a pair of blade-like rami.

This is one of the Chengjiang species that has been reported to have crustacean-like affinities. In the original description *Pectocaris* was erected as the sole member of the Family Pectocarididae and was considered to be a possible branchiopod. Its long carapace is reminiscent of those of *Odaraia alata* from the Burgess Shale and *Vladicaris subtilis* Chlupáč, 1995 from the Lower Cambrian of the Czech Republic. In contrast, Budd's (2002) analysis places *Pectocaris* as a stem-group euarthropod close to the Chengjiang genus *Fuxianhuia*.

P. spatiosa is likely to have been a swimmer. The species is known only from the Chengjiang biota.

Key References Hou & Bergström 1997, Hou 1999, Hou *et al.* 1999.

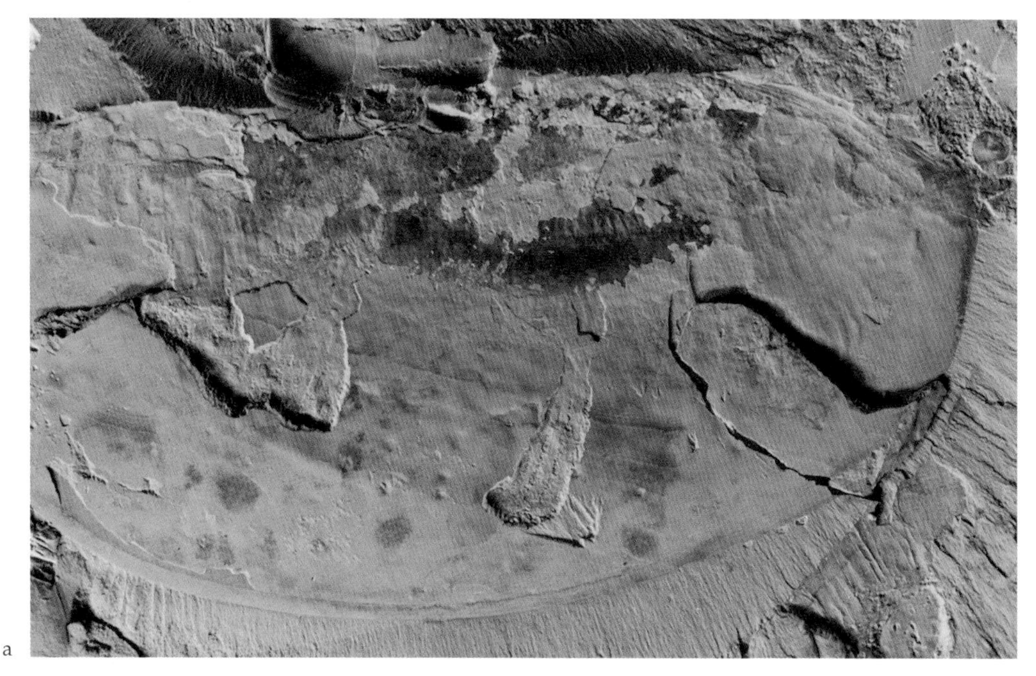

a

b

Figure 16.35 *Pectocaris spatiosa*. (a) Lateral view (NIGPAS 115411), ×1.5; Xiaolantian. (b) Dorsal view (NIGPAS 115377; figured as *Odaraia*? *eurypetala* by Hou & Bergström 1997, Hou *et al*. 1999), ×4.0; Maotianshan.

Genus *Branchiocaris* Briggs, 1976

Branchiocaris? yunnanensis Hou, 1987

This is a fairly common species, but unfortunately only the bivalved carapace is known. Valves are wrinkled, presumably as a result of compaction.

The carapace has a straight dorsal margin to an otherwise subcircular outline, and can be more than 50 mm long and 40 mm high. The anterodorsal and posterodorsal corners of the valves are projected into small processes.

This species may be related to the Burgess Shale species *Branchiocaris pretiosa* (Resser, 1929), in which the appendages are known. The affiliation of *Branchiocaris* is uncertain. The appendage morphology of *B. pretiosa* is reminiscent of branchiopod crustaceans (Hou & Bergström 1997). Cladistic analysis has placed *Branchiocaris* outside the crustaceans and arachnomorphs and close to the Burgess Shale genus *Marrella* Walcott, 1912 and its allies (Wills *et al.* 1998).

B. yunnanensis is known only from the Chengjiang biota.

Key References Hou 1987c, Hou & Bergström 1997, Hou *et al.* 1999.

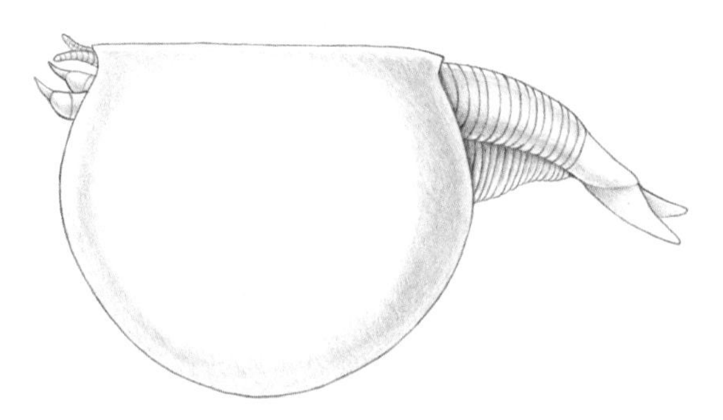

Figure 16.36 Reconstruction of *Branchiocaris? yunnanensis*. The soft part morphology is based on *Branchiocaris pretiosa* (see Briggs *et al.* 1994).

Figure 16.37 *Branchiocaris? yunnanensis*. Lateral view (RCCBYU 10274), ×2.6; Maotianshan.

Genus *Parapaleomerus* Hou, Bergström, Wang, Feng & Chen, 1999

Parapaleomerus sinensis Hou, Bergström, Wang, Feng & Chen, 1999

This species is known from only three specimens, the largest of which is 92 mm long and 90 mm wide. The fossils are dorsoventrally compressed and seen from the dorsal side. No remains of the soft parts of the animal are known.

The dorsal exoskeleton is relatively short and broad, and behind the head its sides slope down from a rounded mid-line. The head shield is short and wide, and lacks any indication of the presence of eyes. The trunk appears to have ten basically similar tergites, of which the last three are less well defined laterally. An eleventh tergite, the telson, is short and has a rounded posterior end.

Parapaleomerus shows similarities to three other rare, monotypic genera, namely the Early Cambrian *Paleomerus* Størmer, 1956 from Sweden, and the Late Cambrian *Strabops* Beecher, 1901 and Late Ordovician *Neostrabops* Caster & Macke, 1952 from the USA. *Parapaleomerus* may be a synonym of *Paleomerus*, although the possible absence in the former, and the presence in the latter, of dorsal eyes, indicates that they may be distinct genera. Dunlop & Selden (1998) regarded *Paleomerus* as a basal chelicerate and perhaps the best model of a primitive arachnomorph. In general design *P. sinensis* also shows a likeness to the Burgess Shale arachnomorphs *Molaria spinifera* and *Emeraldella brocki* (both Walcott 1912), but their exoskeletons end in a very long, narrow telson.

Appendages are not known from *P. sinensis* and other evidence of its ecology is lacking. The species is restricted to the Chengjiang biota.

Key Reference Hou *et al.* 1999.

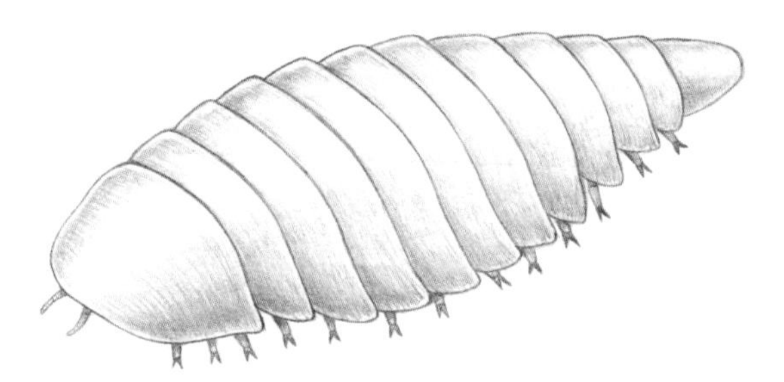

Figure 16.38 Reconstruction of *Parapaleomerus sinensis* (the morphology of the appendages is hypothetical).

Figure 16.39 *Parapaleomerus sinensis*. Dorsal view (NIGPAS 115439), ×1.7; Xiaolantian.

Genus *Naraoia* Walcott, 1912

Naraoia longicaudata Zhang & Hou, 1985

This species is known from over 1,000 specimens. It has a long, smoothly outlined trunk shield which, as in all *Naraoia* species, is effaced. The head shield is sub-semicircular.

Mature specimens can be over 60 mm long, exclusive of antennae. There are at least 22 pairs of basically similar biramous appendages; the first three and (Edgecombe & Ramsköld 1999b) the anterior half of the fourth are in the head region. Distal to the basis, the endopod consists of seven podomeres including a distal claw. At least in the larger appendages the basis and proximal two podomeres each have an endite with spines. The basis is attached to a flexible basal stem (cormus), which enabled leg movement. The exopod is attached to the basis, and at least the proximal part of the first podomere of the endopod. It comprises a long slender proximal segment with long setae, and a narrow distal lobe with bristles. Ventrally, between the antennae, a pair of small rounded structures may represent eyes. The gut is relatively narrow, the cheek diverticulae short, relatively simple and less distinct.

Naraoiids are characterized in part by an unmineralized cuticle, an anteriorly narrowing trunk shield, and the lack of dorsal eyes. Some workers regard them as trilobites (e.g. Fortey & Theron 1994, Wills *et al.* 1998), others do not (e.g. Hou & Bergström 1997, Edgecombe & Ramsköld 1999b). Also, *N. longicaudata* has been considered (Chen *et al.* 1997) distinct enough from the type species of *Naraoia*, *N. compacta* from the Burgess Shale, to erect for it the genus *Misszhouia*.

The spiny endites imply that *N. longicaudata* was at least in part raptorial. However a sediment-ingesting deposit feeding habit has been suggested for it, and/or for *Naraoia spinosa* from Chengjiang (Chen *et al.* 1997, Hou & Bergström 1997, Edgecombe & Ramsköld 1999b, Bergström 2001). Alternatively, the mud within the gut of the Chengjiang naraoiids may have been introduced by turbidity currents (Edgecombe & Ramsköld 1999b), or result from the weathering of phosphate permineralizations of midgut glands (Butterfield 2002). It has also been claimed that Cambrian naraoiids were epibenthic scavengers/predators rather than mud eaters (Vannier & Chen 2002).

N. longicaudata is known only from the Lower Cambrian of Yunnan Province. *Naraoia* also occurs in the Middle Cambrian Kaili Lagerstätte of Guizhou Province (Zhao *et al.* 1999a) and, through *N. compacta*, Canada.

Key References Zhang & Hou 1985, Chen *et al.* 1997, Hou & Bergström 1997, Bergström & Hou 1998, Edgecombe & Ramsköld 1999b, Hou *et al.* 1999, Vannier & Chen 2002.

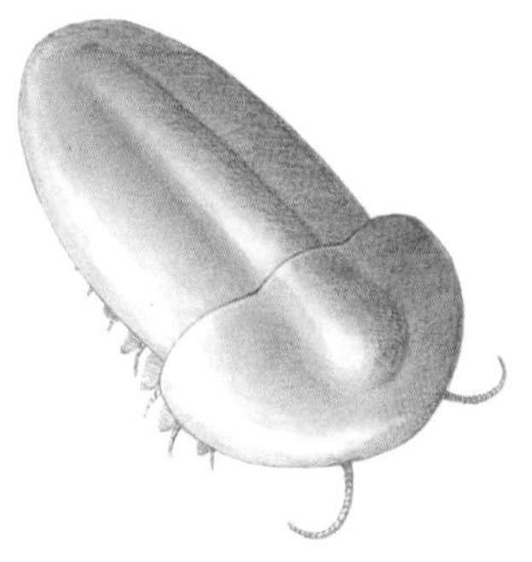

Figure 16.40 Reconstruction of *Naraoia longicaudata*.

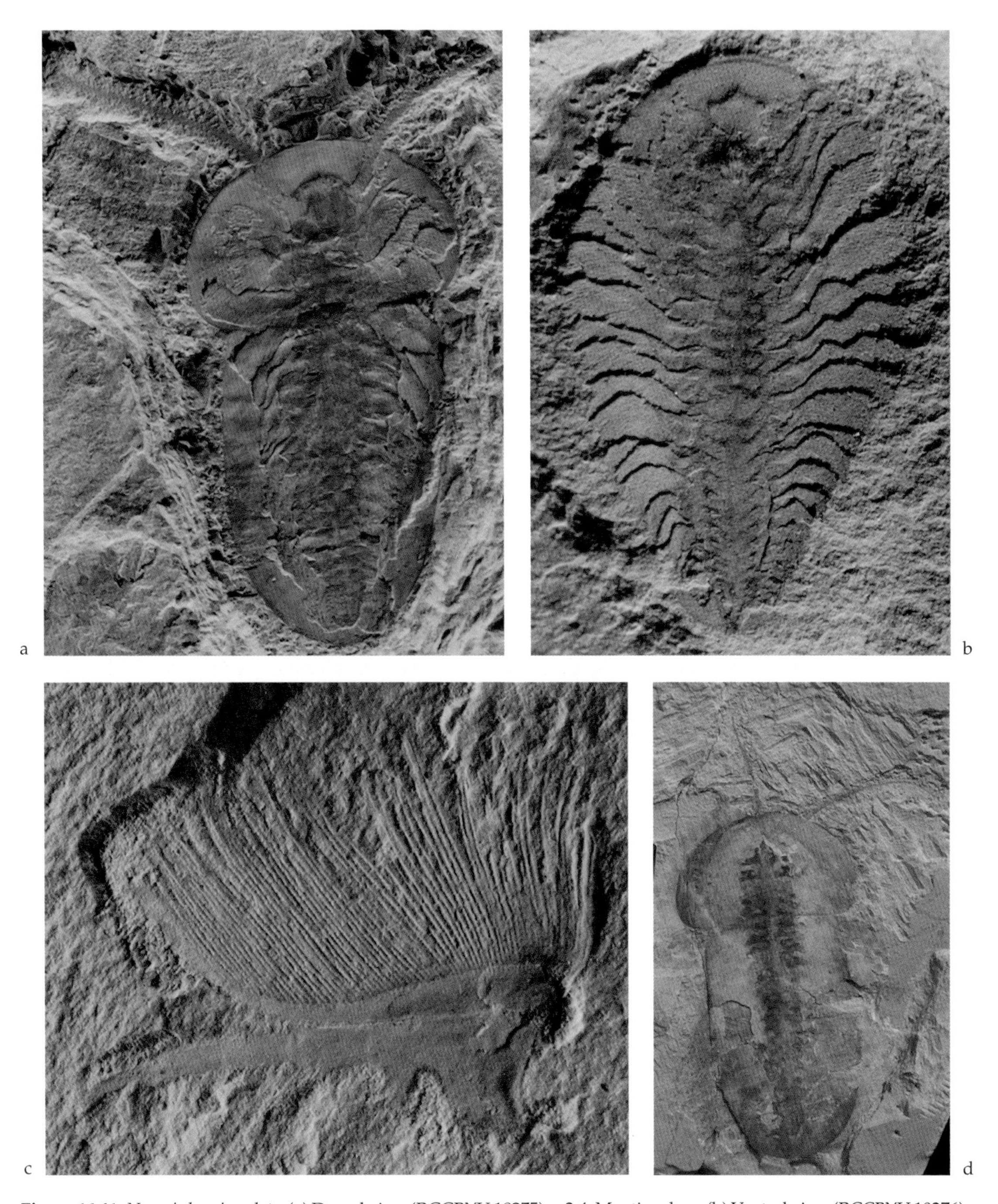

Figure 16.41 *Naraoia longicaudata*. (a) Dorsal view (RCCBYU 10275), × 2.4; Maotianshan. (b) Ventral view (RCCBYU 10276), × 2.7; Maotianshan. (c) Biramous appendage, opened out (NIGPAS 115315), × 5.7; Maotianshan. (d) Dorsal view (RCCBYU 10277), × 0.9; Ma'anshan.

Naraoia spinosa Zhang & Hou, 1985

N. spinosa, like *Naraoia longicaudata*, is known from more than 1,000 specimens. It is a smaller species, with a maximum length of less than 40 mm. The spines indicated by the species name occur on the posterolateral margin of the head shield and lateral and posterolateral margins of the trunk shield, and they are evident in larvae down to 3 mm in length.

As in *N. longicaudata*, there are two types of limbs: uniramous antennae, and biramous appendages of which there are about 15 similar pairs. The biramous appendages have not been described in as much detail as those of *N. longicaudata*, though as in that species the first three pairs, together with the anterior part of the fourth (Edgecombe & Ramsköld 1999b), are positioned beneath the head shield. The alimentary system is distinctly shown in many specimens. It consists of a straight tube extending from the pharynx to the anus, from which arise mostly short, bunch-like diverticulae. In the head region, however, the first pair of diverticular branches gives rise to narrower, longer, ramifying branches.

N. spinosa is readily distinguished from *N. longicaudata* by its much shorter trunk shield; marginal spines that define an embayment posteriorly; much more extensive cheek diverticulae; the lateral deflection, proximally, of the antennae; and an exopod that shows a broader shaft with shorter setae, and a broader distal lobe. In outline it is almost identical to the naraoiid *Liwia plana* from the Lower Cambrian of Poland (Dzik & Lendzion 1988, Hou & Bergström 1997), though the latter differs in having four free thoracic segments.

N. spinosa has been interpreted as either a mainly sediment-ingesting deposit feeder, or a scavenger/predator, as for *N. longicaudata* (see discussion of that species). The different arrangement of the diverticulae of the head region in *N. spinosa* compared to that in *N. longicaudata* has been taken to indicate a more intermittent and opportunistic feeding habit in the former than in the latter (Vannier & Chen 2002).

This species is known only from the Chengjiang fauna.

Key References Zhang & Hou 1985, Chen *et al.* 1997, Hou & Bergström 1997, Bergström & Hou 1998, Hou *et al.* 1999, Bergström 2001, Vannier & Chen 2002.

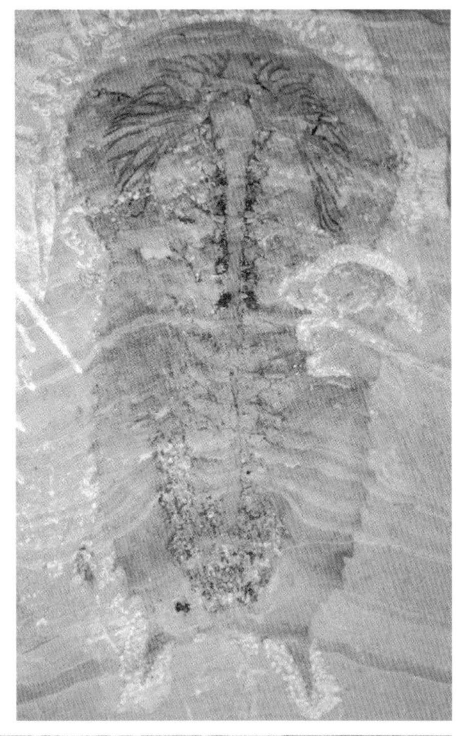

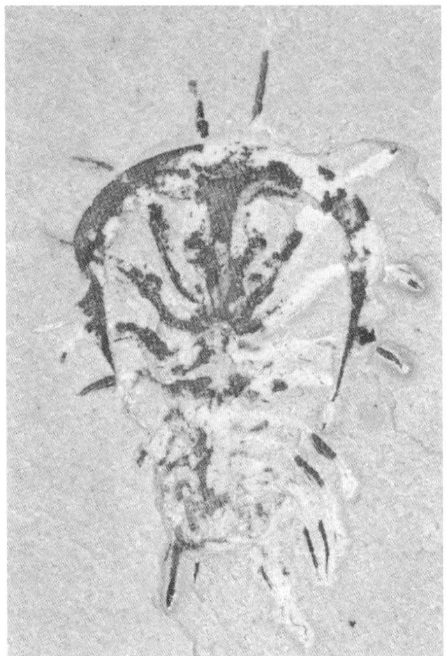

Figure 16.42 *Naraoia spinosa*. (a) Dorsal view (RCCBYU 10279), ×8.0; Maotianshian. (b) Dorsal view (NIGPAS 115385), ×3.7. (c) Detail of head shield (NIGPAS 115385), ×4.9; Maotianshan.

Figure 16.43 *Naraoia*? sp. Dorsal view of a juvenile specimen (RCCBYU 10280), ×15.0; Mafang. Zhang *et al*. (2003) have just referred such "juveniles" to a new, pedomorphic arthropod, *Primicaris*.

Genus *Eoredlichia* Zhang, 1951

Eoredlichia intermedia (Lu, 1940)

E. intermedia is a relatively common trilobite in the Chengjiang biota; the genus lends its name to the *Eoredlichia-Wutingaspis* trilobite Biozone. Antennae, biramous appendages, and supposed traces (dark markings), axially, of the gut and related outgrowths, are known for this species. Also present are caecal features on the cheek area that indicate underlying vessels of the type variously suggested for similar Cambrian trilobites as representing optic nerves or part of a blood circulatory system.

Long, uniramous antennae diverge anteriorly from underneath the head shield. Post-antennal appendages are biramous, with the endopod comprising seven podomeres, the first of which joins proximally with a basis. The basis and the first and second podomeres have spines on their inner surfaces; the seventh podomere is a terminal claw. The exopod consists of a long, broad shaft that bears long setae, and a distal lobe fringed with bristles. It is attached to the entire length of the basis by a hinge joint.

The head shield has a forwardly tapering glabella that is well rounded anteriorly. In front of the neck ring, which is longest medially and shows weak lateral lobes, the first furrow runs backwards and inwards before turning across the central glabellar area, and the second furrow is weaker. There is a preglabellar field posterior to the anterior border. A long cheek spine extends to or beyond the posterior margin of the tail shield. The facial sutures are strongly divergent anteriorly and posteriorly. The hypostome and rostral plate (two ventral sclerites), are attached beneath the anterior glabellar margin (conterminant mode; Fortey 1990). Large crescentic eyes extend posteriorly to the neck furrow. Fifteen laterally spinose segments make up the thorax, with the ninth axial ring supporting a long spine. The tail shield is very small, with two axial rings and a weak third, and two fused segments and associated furrows on the flank area. The cuticle is variously finely pustulose.

Redlichioids of the *Eoredlichia* type have been considered to be the sister group of a group comprising all "higher" (non-olenelloid) trilobites (Fortey 1990). *Eoredlichia* lived on or close to the sea-bottom. One study has suggested that the long setae on the proximal part of the exopod may have been used for filtering food (Shu *et al.* 1995a). Another analysis, which assessed feeding habits in trilobites in general, argued for *Eoredlichia* being a predator based on the nature of its stout spinose limb bases and its braced hypostome, and for this feeding mode to be basal to all non-olenelloid trilobites (Fortey & Owens 1999).

This species has been recorded from many Lower Cambrian localities in Yunnan Province.

Key References Lu 1940, Zhang 1951, Shu *et al.* 1995a, Ramsköld & Edgecombe 1996.

Figure 16.44 *Eoredlichia intermedia*. (a) Dorsal view (RCCBYU 10281), ×0.8; Ma'anshan. (b) Dorsal view (RCCBYU 10282), ×1.7; Maotianshan.

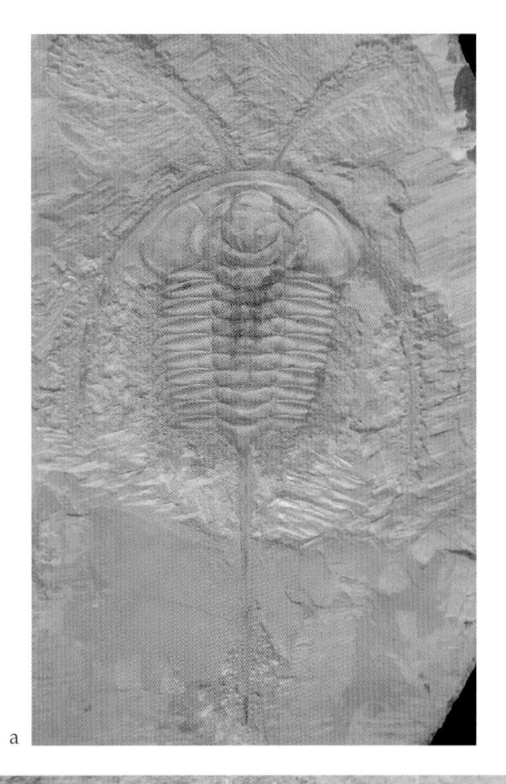
a

b

Genus *Kuanyangia* Hupé, 1953

Kuanyangia pustulosa (Lu, 1941)

This species is relatively rare compared with the other trilobites in the Chengjiang fauna. Specimens with limbs have been found.

The uniramous antennae attach under the frontal lobe of the glabella; each is stout proximally, beyond which there are about 20 annuli, though the most distal part is unknown. Four cephalic exopods have been identified, together with, in both the cephalic and thoracic limbs, a large basis that is finely serrated medially and ventrally. Five or six cylindrical podomeres extend beyond the basis, the terminal one bearing a few short spines. Two segments form the large, blade-like exopod, fringed with narrow, flat setae posteriorly and distally.

The head shield is sub-semicircular in outline. The glabella is conical; it has three distinct furrows, all of which are deepest as they run obliquely backwards, before they continue more weakly across the central glabellar area. The neck ring bends forwards laterally into a lobe. There is a short preglabellar field behind a convex anterior border. A medium-sized eye lobe is sited at the mid-length of the cheek, and the eye ridge is strong. Facial sutures diverge anteriorly and run outward and backward posteriorly. There is a relatively short cheek spine. The thorax comprises 16 segments, the axial rings of which have a median node and a lateral lobe, and there are well-developed lateral marginal spines. The tail shield is very small with two or three axial rings. Pustules cover much of the cuticle.

K. pustulosa is a redlichioid trilobite, benthic in habit, and is confined to the *Eoredlichia-Wutingaspis* trilobite Biozone of Yunnan Province.

Key References Lu 1941, Hupé 1953, Hou & Bergström 1997.

Figure 16.45 *Kuanyangia pustulosa*. (a) Dorsal view, part (NIGPAS 115407a), ×1.6; Maotianshan. (b) Dorsal view, counterpart (NIGPAS 115407b), ×1.6.

Genus *Yunnanocephalus* Kobayashi, 1936

Yunnanocephalus yunnanensis (Mansuy, 1912)

Y. yunnanensis is a moderately common trilobite in the Chengjiang biota. Antennae and biramous appendages have been prepared out in a few specimens. Cheek caecae, reflecting an underlying network of vessels or nerves, are also known. The antennae are subparallel proximally from their origins to the margin of the head shield, in front of which they diverge strongly and taper. The biramous appendages (present in both the cephalon and thorax) are not well enough preserved to allow detailed morphological interpretation.

The head shield is well rounded in outline. The glabella tapers anteriorly and has straight sides, a bluntly rounded frontal lobe, and weak or very weak furrows. The neck ring comprises a broad band. The preglabellar field is slightly longer than the anterior border. The eye is relatively small, the eye ridge often bilobate. Facial sutures are subparallel to weakly converging anteriorly; posteriorly they run obliquely backwards to bisect a small inter-cheek spine inside a rounded cheek margin. The hypostome has a conterminant attachment style, and has been inferred to connect to the ventral continuation of the anterior border by means of a plectrum-like rostral plate. Fourteen segments make up the thorax, with all but the most posterior axial rings showing lateral lobes, and with short, stout lateral marginal spines. A median node is present on most axial rings. The tail shield is very small and largely consists of a subquadrate axis that has two incomplete axial rings and a ventral boss below its posterior tip. The flank area is tiny and subtriangular.

Y. yunnanensis belongs, like the Chengjiang species *Eoredlichia intermedia* and *Kunyangia pustulosa*, to the Redlichioidea, a relatively primitive group of trilobites. All probably lived on or near the sediment/water interface. *Yunnanocephalus* is known from various localities in Yunnan Province.

Key References Mansuy 1912, Kobayashi 1936, Shu *et al.* 1995a.

Figure 16.46 *Yunnanocephalus yunnanensis*. (a) Dorsal view (RCCBYU 10283), ×9.0; Maotianshan. (b) Dorsal view of three specimens (RCCBYU 10284), ×3.5; Xiaolantian.

Genus *Retifacies* Hou, Chen & Lu, 1989

Retifacies abnormalis Hou, Chen & Lu, 1989

This rare species attained considerable size. Individuals with a total length, including the antennae and tail, of 12 cm are known, but most specimens are much smaller. Behind the short head shield there are ten, overlapping thoracic tergites and a large tail shield, articulated to which there is a long, segmented tail. All of the tergites have an irregular polygonal mesh-like surface ornament not seen in any other Chengjiang arthropod.

Ventrally, near the prominent hypostome, club-shaped eyes are set on simple stalks, projecting beyond which are long setiferous antennae. Behind the antennae there are about 18 pairs of biramous appendages, three in the head, ten in the thorax and approximately five in the pygidium. The cephalic biramous appendages are similar to the trunk appendages, but closer together. The inner branch of each appendage consists of podomeres with a multispinose inner edge. The exopod is a crescent-shaped proximal flap fringed by a splay of about 20 long lamellar setae that are edged with bristles distally.

Opinions differ concerning the affinities of the trilobite-like *Retifacies*. Comparisons have been drawn with helmetiids and naraoiids (Delle Cave & Simonetta 1991, Hou & Bergström 1997). In contrast, a study of Cambrian arachnomorphs concluded that a naraoiid-*Retifacies* grouping was paraphyletic (Edgecombe & Ramsköld 1999b). Some of the analyses of the latter authors resolve *Retifacies* close to Walcott's (1911a, 1912) Burgess Shale genera *Sidneyia* and *Emeraldella*.

Like many other contemporary arthropods, *R. abnormalis* probably lived crawling on the sea-bottom. Although it does not seem to have any specialized head appendages, the serrated inner edge of each leg suggests that it may have been predatorial or a scavenger. The report of sediment infilling part of the intestine in one specimen (Hou & Bergström 1997) suggests other feeding strategies.

Retifacies is known only from the Chengjiang biota.

Key References Hou *et al.* 1989, Chen & Zhou 1997, Hou & Bergström 1997, Edgecombe & Ramsköld 1999b, Hou *et al.* 1999.

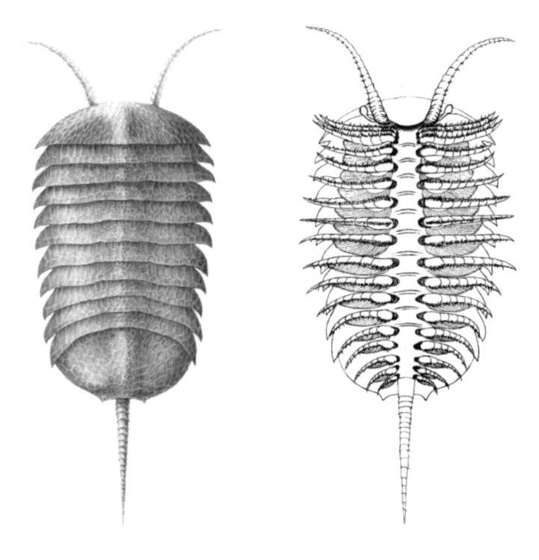

Figure 16.47 Reconstruction of *Retifacies abnormalis* in dorsal and ventral views (after Hou & Bergström 1997).

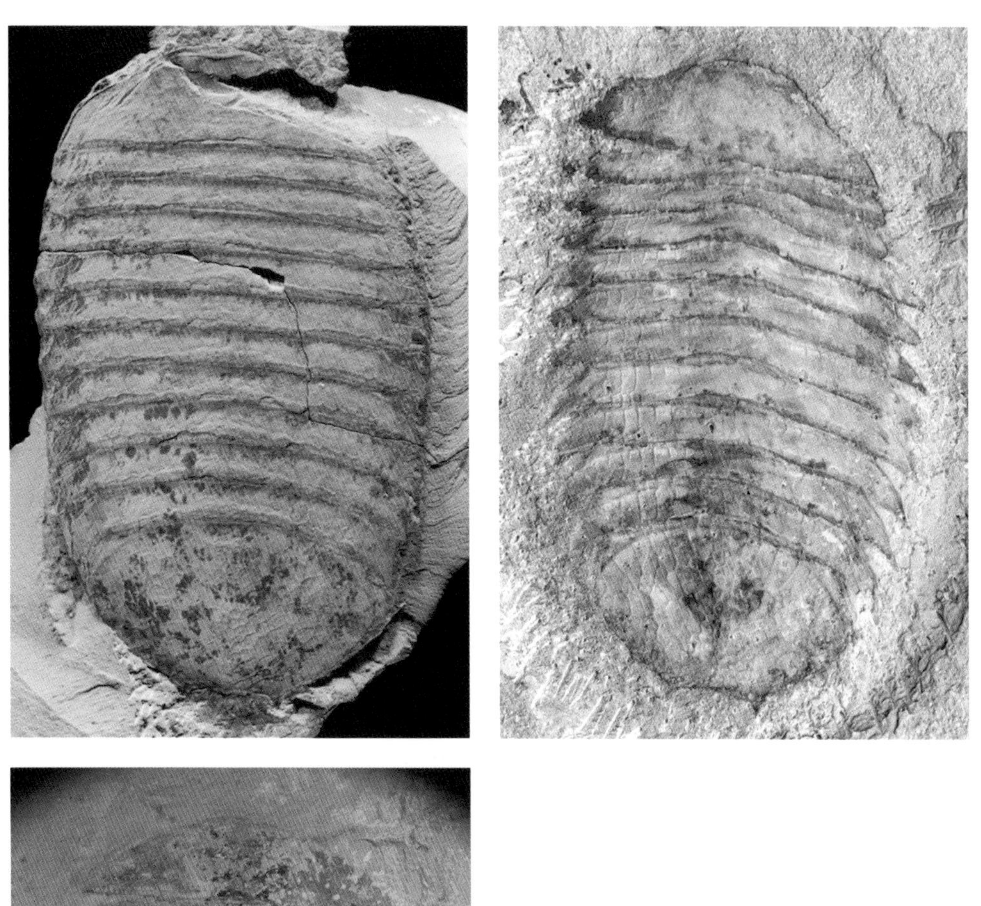

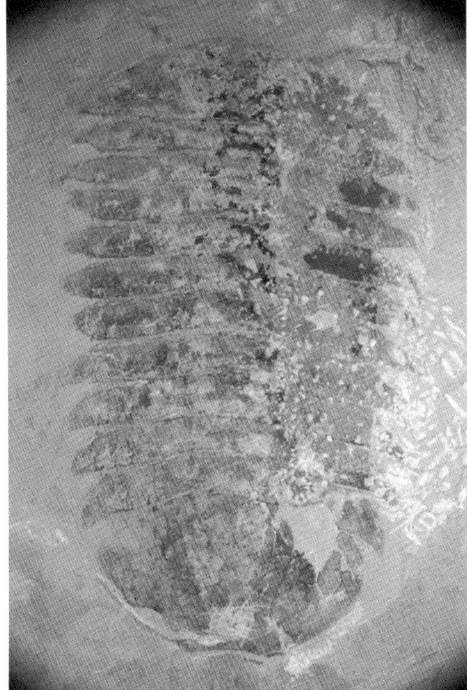

Figure 16.48 *Retifacies abnormalis*. (a) Dorsal view (NIGPAS 115390), ×1.4; Maotianshan. (b) Dorsal view (RCCBYU 10285), ×1.5; Mafang. (c) Ventral view (NIGPAS 115388b), ×2.3; Maotianshan.

Genus *Squamacula* Hou & Bergström, 1997

Squamacula clypeata Hou & Bergström, 1997

This species is known from only a few, dorsoventrally compressed specimens, up to about 1 cm long. The dorsal exoskeleton and parts of the gut and appendages are preserved.

The dorsal exoskeleton is smooth, broad and vaulted, but there is no discrete axial region. Behind the short head shield is a trunk with ten overlapping tergites, each extended into small spines laterally, and a tiny terminal tergite. The multiannulate antennae are long. The other appendages have an inner, "walking" branch and an outer branch with setae, but their detailed structure is unknown.

S. clypeata has been tentatively regarded as a retifaciid (Hou & Bergström 1997). The well-developed antennae, the lack of dorsal eyes, the short and broad head shield, the shape of the tergites and the lack of a terminal spine are characters in common with the Chengjiang associate *Retifacies abnormalis*.

S. clypeata has a somewhat flattened shape and may have inhabited the sea-bottom. Its gut apparently contains silt, and it may have had sediment-ingesting habits (Hou & Bergström 1997).

This taxon is the only recorded species of *Squamacula* and is unknown outside the Chengjiang fauna.

Key References Hou & Bergström 1997, Hou *et al.* 1999.

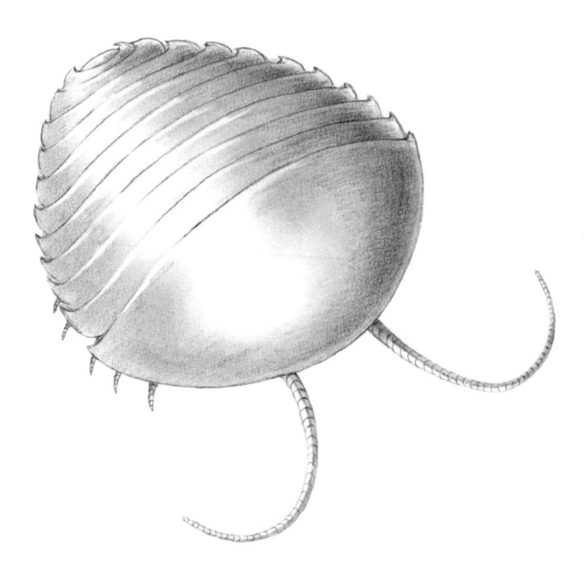

Figure 16.49 Reconstruction of *Squamacula clypeata*.

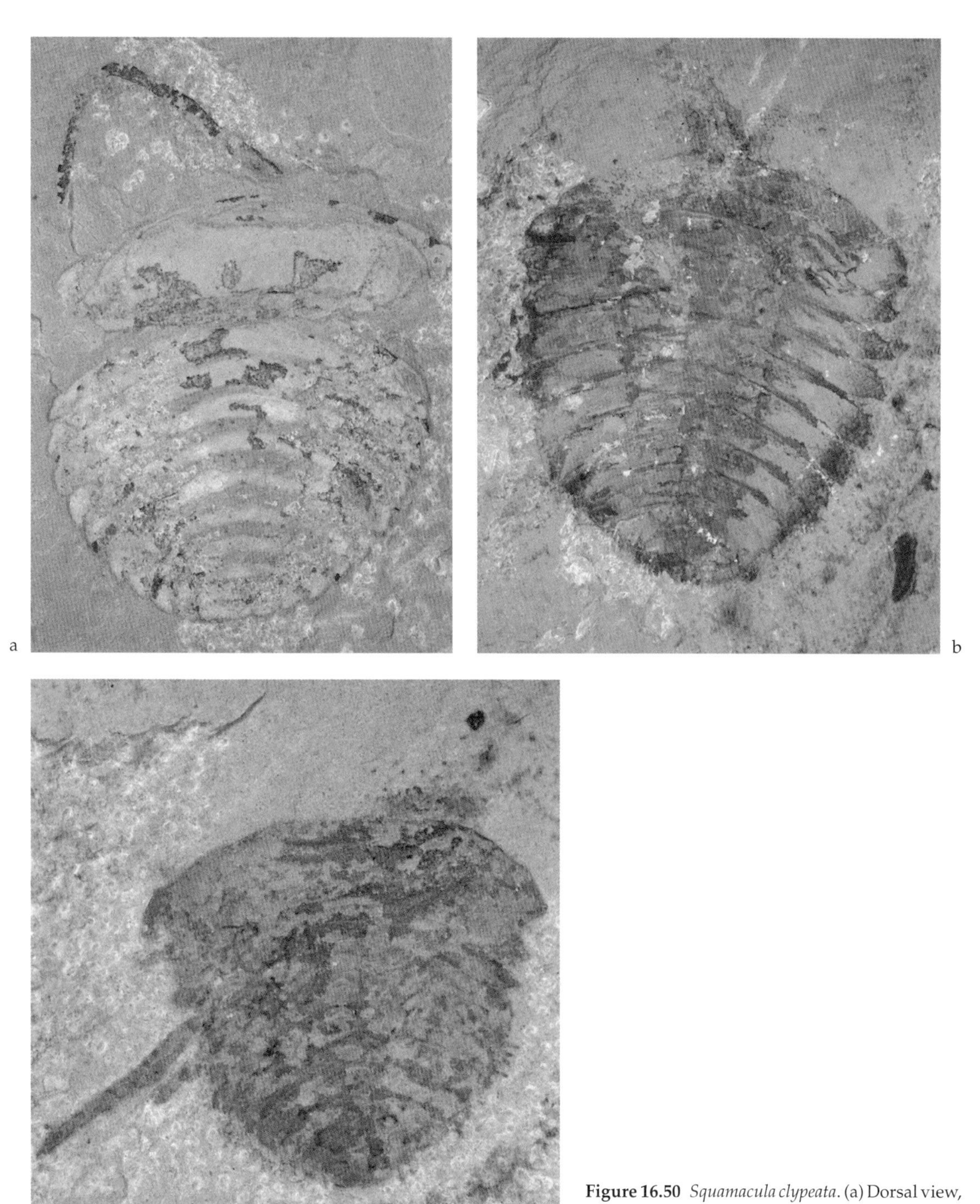

Figure 16.50 *Squamacula clypeata*. (a) Dorsal view, counterpart (NIGPAS 115393), ×7.0; Maotianshan. (b) Dorsal view (RCCBYU 10286), ×4.1; Xiaolantian. (c) Dorsal view (NIGPAS 115494), × 7.8; Xiaolantian.

Genus Kuamaia Hou, 1987

Kuamaia lata Hou, 1987

This species is known from more than 100 fossils, the largest being at least 10 cm long excluding appendages. The unmineralized cuticle is preserved as a thin, white or purplish-red film.

The relatively flat dorsal exoskeleton is differentiated into a head shield, a thorax with seven tergites, and a pygidium. Parts of the exoskeleton are fused, though opinions vary on features affected and its extent (see Hou & Bergström 1997, Edgecombe & Ramsköld 1999b). A pair of low relief bulges forward on the head shield mark the position of the eyes, which are stalked and originate ventrally. *Kuamaia*, its Chengjiang associates *Saperion* and *Skioldia* and the Burgess Shale genus *Helmetia* Walcott, 1918 have similar types of eyes and an anterior sclerite (rostral plate) that curves from dorsal to ventral at the front of the animal. The tail of *K. lata* is similar to that of *Helmetia expansa* Walcott, 1918 in having a terminal spine and two pairs of lateral spines.

On the underside, just behind the anterior sclerite, there is a trilobite-type hypostome. The gut is identified as a narrow dark band. The antennae are simple, consisting of uniform, setae-bearing annuli. The basal part (basipod) of the biramous limb is large and spinose. The endopod consists of simple spinose podomeres. The exopod has a petal-shaped proximal element with long lamellar setae extending from one edge and a broad element, seemingly divided into two, along the other edge.

Hou & Bergström (1997) placed *Kuamaia* in the Family Helmetiidae, and identified the monospecific Chengjiang genus *Rhombicalvaria* Hou, 1987 as a possible synonym. That *Kuamaia* is allied to a group of arthropods that includes *Helmetia*, *Skioldia*, *Saperion* and the Burgess Shale *Tegopelte* Simonetta & Delle Cave, 1975 has been endorsed in a cladistic analysis of Cambrian arachnomorphs (Edgecombe & Ramsköld 1999b). *Helmetia* differs from the Chinese genera in having a transverse frontal margin with a prominent pair of lateral spines.

The body shape of *K. lata* suggests that it was a benthic animal. The spiny inner side of its legs indicates that it was possibly carnivorous.

This species is known only from the Chengjiang fauna.

Key References Hou 1987b, Hou & Bergström 1997, Bergström & Hou 1998, Edgecombe & Ramsköld 1999b, Hou *et al.* 1999.

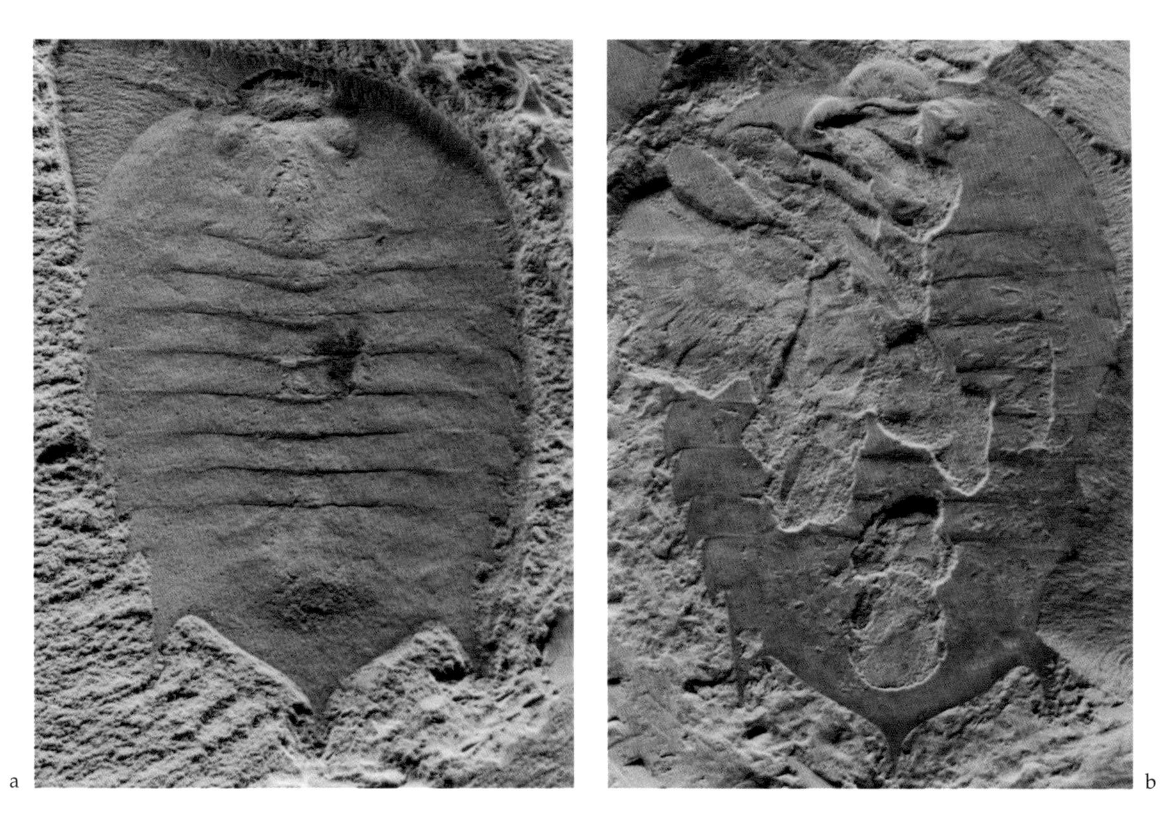

Figure 16.51 *Kuamaia lata*. (a) Dorsal view (NIGPAS 115400), ×2.3; Maotianshan. (b) Dorsal view (NIGPAS 115318), ×2.0; Maotianshan.

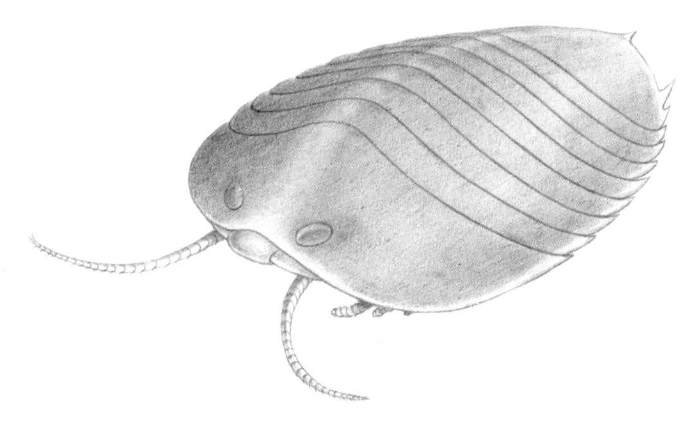

Figure 16.52 Reconstruction of *Kuamaia lata* (after Hou & Bergström 1997).

Genus *Skioldia* Hou & Bergström, 1997

Skioldia aldna Hou & Bergström, 1997

This species is known from only a few specimens. Some individuals reach a length of more than 10 cm. The exoskeleton was unmineralized and is preserved flattened.

The dorsal exoskeleton has a broadly oval shape and, though clearly segmented, is fused into a single shield. A total of about 13 segments are visible, the boundaries of which are more distinct in the axial and middle (thoracic) parts of the body than in the lateral and posterior portions. The eyes are accommodated in exoskeletal bulges and occur fairly close to the anterior sclerite (rostral plate), behind which there is a hypostome. Numerous minute spines occur at the margins of the dorsal shield, especially on the tail. Behind the multiannulate antennae each segment has paired biramous appendages, as seen impressed on the surface of the exoskeleton. Long lamellar setae are discernible on the exopods, but these limbs await detailed study.

The Family Skioldiidae is based on the single species, *S. aldna*. Cladistic analysis of Cambrian arachnomorphs resolves *Skioldia* close to the Chengjiang genera *Saperion* and *Kuamaia* and the Burgess Shale genera *Helmetia* and *Tegopelte* (Edgecombe & Ramsköld 1999b). The shields of *Skioldia* and *Kuamaia* differ in shape, and the marginal spines in *Skioldia* are much smaller and segmentation is more indistinct. *Skioldia* appears to have taken the fusion of the dorsal skeletal elements further than in *Kuamaia*, and is intermediate between *Kuamaia* and *Saperion* in this respect.

This animal would probably have had much the same benthic lifestyle as *Kuamaia*. *S. aldna* is known only from the Chengjiang fauna.

Key References Hou & Bergström 1997, Edgecombe & Ramsköld 1999b, Hou *et al.* 1999.

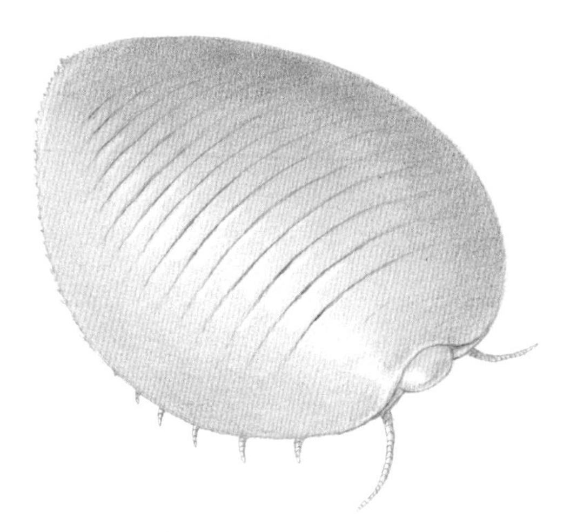

Figure 16.53 Reconstruction of *Skioldia aldna* (the morphology of the endopods is hypothetical).

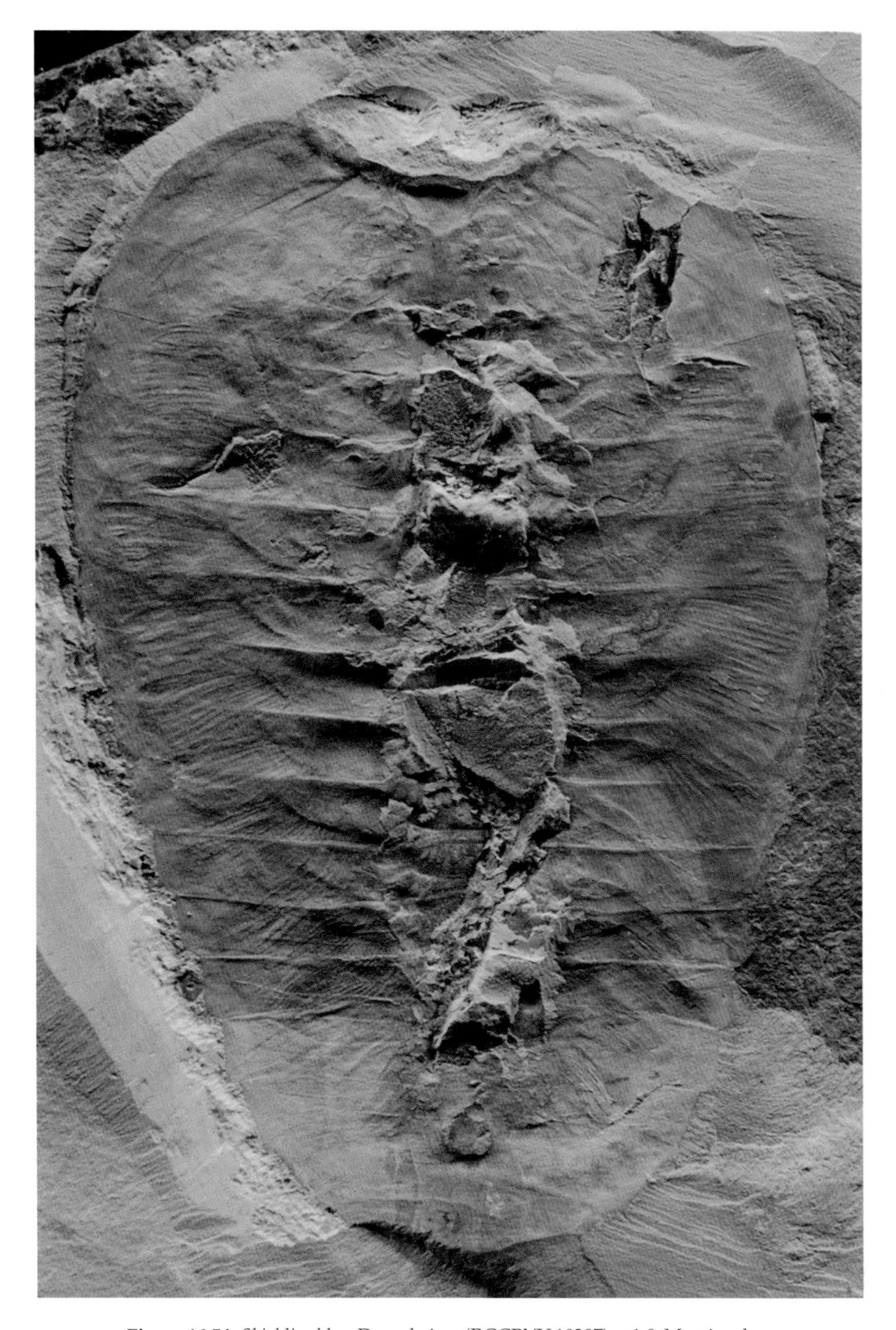

Figure 16.54 *Skioldia aldna*. Dorsal view (RCCBYU 10287), × 1.9; Maotianshan.

Genus *Saperion* Hou, Ramsköld & Bergström, 1991

Saperion glumaceum Hou, Ramsköld & Bergström, 1991

S. glumaceum is rare. Specimens range from about 2 cm to over 15 cm in length and are preserved as flattened impressions with virtually no relief. The exoskeleton lacked mineralization.

This species represents a more complete example of the trend—seen to a lesser degree in *Kuamaia* and *Skioldia* from Chengjiang—to fuse the components of the dorsal exoskeleton. The anterior part of the single shield lacks segment boundary lines, and thereby indicates the extent of the head. Behind, weakly impressed furrows outline some 20 segments, but they do not reach the lateral margins of the shield. It is hardly possible to distinguish a boundary to define "thorax" and "pygidium". *Saperion* has a more elongate outline than *Kuamaia* and *Skioldia* and it lacks marginal spines.

Anteriorly in the head a pair of antennae flanks the rostral plate and hypostome. Ventral eyes are indicated by a weak bulge in the dorsal exoskeleton and are similar to those in *Kuamaia*, and apparently *Sinoburius*, from Chengjiang (Edgecombe & Ramsköld 1999b). The biramous appendages of the trunk have a well-developed basipod from which stems an inner, "walking" branch of simple podomeres and an exopod consisting of a proximal lobe with numerous long lamellar setae and an unadorned distal lobe. The biramous appendage of *Kuamaia* appears to be fundamentally similar except that the distal lobe of the exopod is cleft into two parts. The alliance of *Saperion* (Family Saperiidae) with *Skioldia*, *Kuamaia* and similar Cambrian arachnomorphs is supported by cladistic analysis (Edgecombe & Ramsköld 1999b).

S. glumaceum probably led much the same lifestyle, on the sea-bottom, as species of *Kuamaia* and *Skioldia*. It shows no obvious morphological features suggestive of carnivorous habits.

The species is known only from the Chengjiang biota.

Key References Hou *et al.* 1991, Ramsköld *et al.* 1996, Chen & Zhou 1997, Hou & Bergström 1997, Edgecombe & Ramsköld 1999b, Hou *et al.* 1999.

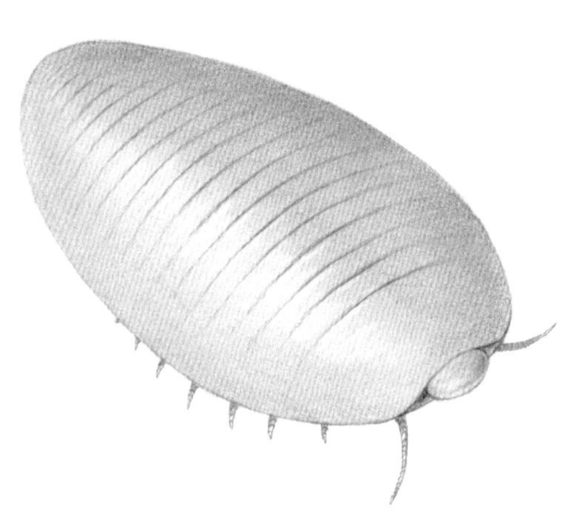

Figure 16.55 Reconstruction of *Saperion glumaceum*.

Figure 16.56 *Saperion glumaceum*. Dorsal view (NIGPAS 115289), ×7.3; Jianbaobaoshan, near Dapotou.

Genus *Cindarella* Chen, Ramsköld, Edgecombe & Zhou *in* Chen *et al.*, 1996

Cindarella eucalla Chen, Ramsköld, Edgecombe & Zhou *in* Chen *et al.*, 1996

This species has been reported from only a few, dorsoventrally compressed specimens. The exoskeleton lacked mineralization and is about 11 cm long excluding appendages.

As in the closely related Chengjiang genus *Xandarella*, *C. eucalla* has a large semicircular "head shield". In *Cindarella* this "shield" belongs to only the first five (antennal plus four post-antennal) segments, and its posterior part is extended as a (carapace) fold to cover an additional six anterior trunk tergites (Ramsköld *et al.* 1997). The paired ventral eyes extend anterolaterally beyond the margin of the "head shield", as do the long antennae. The trunk has 15–17 tergites in addition to those overlapped by the posterior extension of the "head shield", with a median spine on the posterior-most tergites. The gut can be traced from near the hypostome to close to the posterior margin of the terminal tergite.

C. eucalla shows a decoupling of tergites and segments in the trunk (Ramsköld *et al.* 1997). The anterior six trunk tergites correspond to single segments and, like the four post-antennal head segments, each carries a pair of biramous limbs. Behind, the other trunk tergites each cover more than one segment (as expressed by limb pairs), with a lack of correspondence between tergite boundaries and segment boundaries. The number of appendages per tergite increases progressively from one to about four posteriorly, so that the 15 or so multisegmental tergites behind the posterior extension of the "head shield" correspond to some 20 limb pairs and the pygidium houses even more. In this way, there is a total of 35–40 limb pairs. The situation differs from that in the otherwise basically similar *Xandarella*, where limbs seem to correspond to true trunk segments, but sets of segments share a tergite.

Cindarella and *Xandarella* formed the basis of the Xandarellida (Chen *et al.* 1996, Hou & Bergström 1997, Ramsköld *et al.* 1997), a group closely related to the Chengjiang arachnomorph *Sinoburius* Hou *et al.*, 1991 (Hou & Bergström 1997, Edgecombe & Ramsköld 1999b). *Cindarella* differs from *Xandarella* in the number of cephalic segments and in having eyes that extend laterally from under the margin of the head shield. The Chengjiang species *Almenia spinosa* was regarded as synonymous with *C. eucalla* by Edgecombe & Ramsköld (1999b).

In common with the Chenjiang genus *Saperion*, *Cindarella* lacks major endites on the endopods and gnathobases have not been observed on the limb base. There is no evidence of raptorial habits.

C. eucalla is known only from the Chengjiang fauna.

Key References Chen *et al.* 1996, Chen & Zhou 1997, Ramsköld *et al.* 1997, Edgecombe & Ramsköld 1999b.

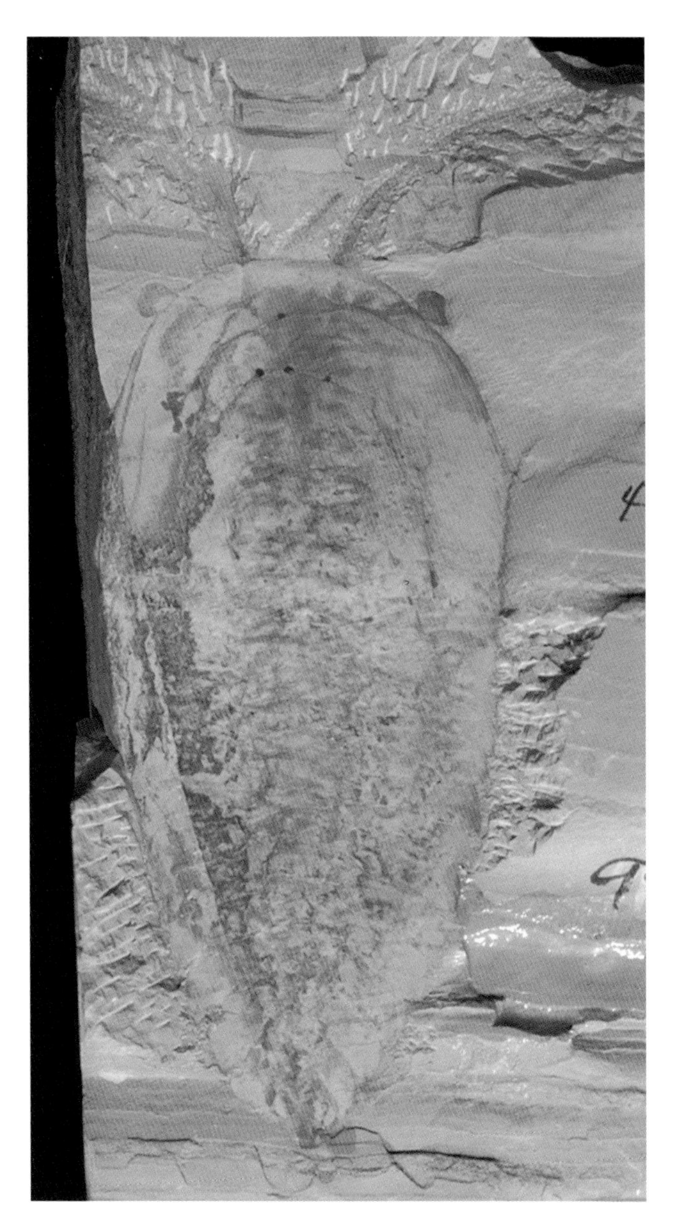

Figure 16.57 *Cindarella eucalla*. Dorsal view (RCCBYU 10288), × 1.0; Mafang.

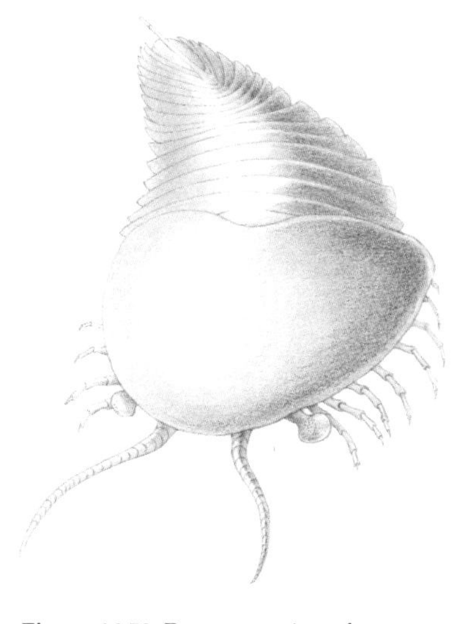

Figure 16.58 Reconstruction of *Cindarella eucalla*.

Genus *Xandarella* Hou, Ramsköld & Bergström, 1991

Xandarella spectaculum Hou, Ramsköld & Bergström, 1991

This is a rare species, known from well-preserved dorsoventrally flattened fossils, some of which have the gut in relief. Some complete specimens are 50–55 mm long.

The head has six pairs of biramous appendages behind multiannulate antennae flanking a large hypostome. The eyes originate ventrally and are set to see through a round, slightly raised opening in the dorsal part of the exoskeleton. The animal possibly looked upwards, rather than horizontally through a narrow fissure as in early trilobites. A fissure extends from each eye to the lateral margin of the head shield. In the closely related *Cindarella* the eyes project out on a stalk from under the border of the shield. The situation in *Xandarella* appears to be one example of a convergent pattern in arthropods, where eyes have attained a more dorsal position. The head shield is extended posteriorly to overlap the anterior part of the trunk, where there is a small axial tergite.

Behind the small tergite are a further ten segmental tergites and a posterior tergite bearing an axial spine. Surprisingly, there are at least 32 pairs of appendages corresponding to these tergites. Each endopod consists of many podomeres, some with small spines. Lamellar setae are attached to the distal podomeres of the exopods. In contrast to *Cindarella*, the limbs in *Xandarella* seem to correspond to true body segments and sets of segments share a tergite.

X. spectaculum is trilobite-like but differs from trilobites in several characters, such as the morphology of the eye and (notwithstanding the systematic position of naraoiids) the lack of calcification of the exoskeleton. The xandarellids and *Sinoburius* of the Chengjiang biota are closely related arachnomorphs (Hou & Bergström 1997, Ramsköld *et al.* 1997, Edgecombe & Ramsköld 1999b).

The gut of *X. spectaculum* is said to contain sediment, thereby suggesting that the animal engaged in deposit feeding (Hou & Bergström 1997).

Xandarella is known from a single species that is currently found only in the Cambrian of Yunnan Province.

Key References Hou *et al.* 1991, Chen & Zhou 1997, Hou & Bergström 1997, Ramsköld *et al.* 1997, Bergström & Hou 1998, Edgecombe & Ramsköld 1999b, Hou *et al.* 1999.

Figure 16.59 Reconstruction of *Xandarella spectaculum*.

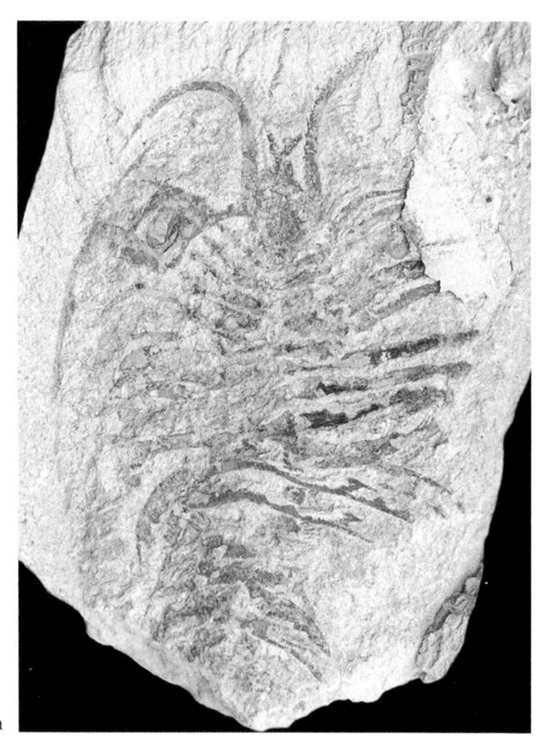

a

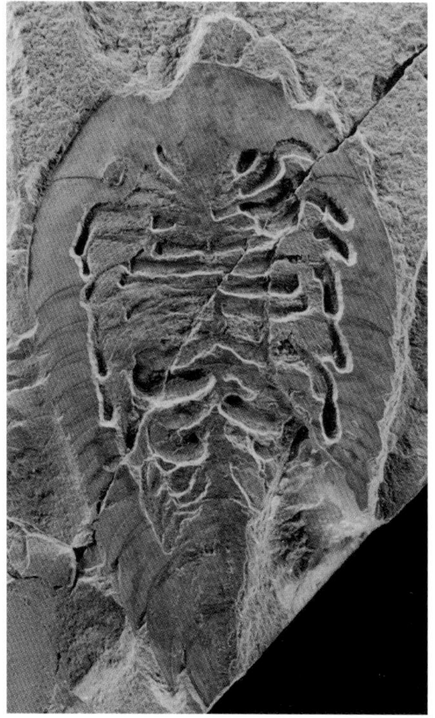

b

c

Figure 16.60 *Xandarella spectaculum*. (a) Ventral view (RCCBYU 10191), ×2.5; Mafang. (b) Dorsal view (NIGPAS 115286a), ×2.1; Jianbaobaoshan, near Dapotou. (c) Dorsal view (NIGPAS 115285), ×1.7; Maotianshan.

Genus *Sinoburius* Hou, Ramsköld & Bergström, 1991

Sinoburius lunaris Hou, Ramsköld & Bergström, 1991

Only about five, flattened specimens of this rare Chengjiang species have been reported. The exoskeleton lacked mineralization, and its general appearance is reminiscent of trilobites.

Adults are small, the largest being only about 1.2 cm long. *S. lunaris* resembles the Chengjiang species *Xandarella spectaculum* in having a broad head shield succeeded by a small tergite and in the lateral spines and an axial spine on the pygidium-like posterior. The eyes, which seem to be ventrally based, stalked, and have dorsal sight through openings in the exoskeleton in both these genera, are large and elongate oval in *S. lunaris*. Opinions differ about the possible presence of a slit extending (as in *Xandarella*) from each eye to the lateral margin of the head shield in *S. lunaris* (Chen *et al.* 1996, Hou & Bergström 1997, Edgecombe & Ramsköld 1999b). Behind the simple antennae, there are at least four pairs of notably long, laterally directed biramous appendages in the head and a pair of biramous appendages under each thoracic tergite (Hou & Bergström 1997).

The Family Sinoburiidae is based on the only known species of *Sinoburius*. Commentators agree that *Sinoburius* is closely related to the Chengjiang xandarellids *Xandarella* and *Cindarella* (Hou & Bergström 1997, Ramsköld *et al.* 1997, Edgecombe & Ramsköld 1999b). In contrast to *Xandarella* and *Cindarella*, in *Sinoburius* there is only one pair of appendages in each trunk tergite.

This is one of the many Chengjiang arthropods in which the gut is apparently stuffed with silt, so most likely it was a benthic deposit feeder, a lifestyle that seems consistent with its overall morphology (Hou & Bergström 1997).

S. lunaris is known only from the Chengjiang biota.

Key References Hou *et al.* 1991, Chen *et al.* 1996, Chen & Zhou 1997, Hou & Bergström 1997, Ramsköld *et al.* 1997, Edgecombe & Ramsköld 1999b, Hou *et al.* 1999.

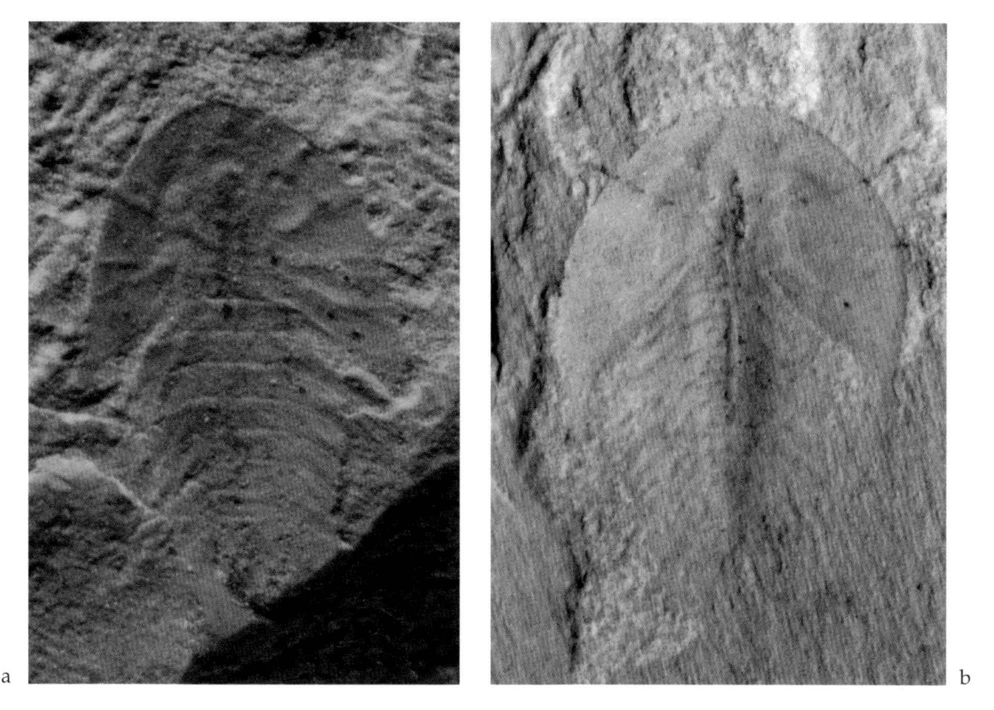

Figure 16.61 *Sinoburius lunaris*. (a) Dorsal view, counterpart (NIGPAS 115287), × 8.3; Maotianshan. (b) Dorsal view (NIGPAS 115288), × 12.0; Maotianshan.

Figure 16.62 Reconstruction of *Sinoburius lunaris*.

Genus *Acanthomeridion* Hou, Chen & Lu, 1989

Acanthomeridion serratum Hou, Chen & Lu, 1989

This is a rare species, originally described from eight specimens, all of which are dorsoventrally flattened.

A. serratum is up to 35 mm long and almost parallel-sided. Its head shield has a rounded anterior margin. One of its distinctive features is that it appears to have free cheeks, separated from the rest of the head shield by a suture like that in most trilobites. The trunk has 11 smooth, well-defined tergites that extend laterally to form posteriorly directed spines that are especially long in the posterior tergites. The posterior-most tergite has a marked medial furrow housing a long narrow spine. The appendages and soft parts of the animal are unknown.

Because of the lack of information on its appendages, and since its morphology is in general quite different from other arthropods, the affinity of this species is not clear. A family and order was established on the basis of the monotypic *Acanthomeridion* (Hou & Bergström 1997).

The shape of its body indicates that *A. serratum* probably lived on the sea-bottom. The lack of knowledge of its limbs makes it difficult to interpret its mode of feeding.

Acanthomeridion is unknown outside the Chengjiang biota.

Key References Hou *et al.* 1989, Hou & Bergström 1997, Hou *et al.* 1999.

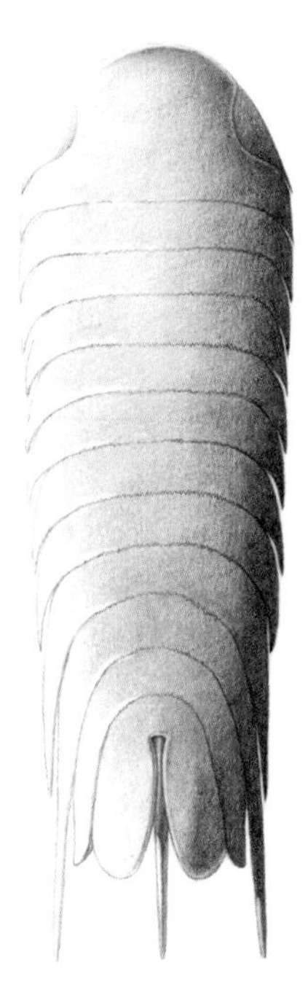

Figure 16.63 Reconstruction of *Acanthomeridion serratum* (after Hou & Bergström 1997).

Figure 16.64 *Acanthomeridion serratum*. Dorsal view (RCCBYU 10290), ×3.9; Maotianshan.

17 PHYLUM BRACHIOPODA

The brachiopods are benthic, marine animals with a shell consisting of two valves, termed dorsal and ventral; the ventral valve is usually the slightly larger of the pair. Attachment to the seafloor is often by means of a fleshy stalk or pedicle that extends from the posterior end of the shell, and brachiopods gain their food from suspended organic matter in the water using a delicate filtering organ termed the lophophore. All the brachiopods from the Chengjiang biota are relatively primitive, lacking hingement structures to operate the valves, which are held together in these forms by internal muscle systems. Two subphyla are represented in the Chengjiang biota: the Linguliformea, in which the shell is chitinophosphatic, and the Craniiformea, which have calcareous shells and in which the pedicle is absent or reduced (e.g. *Heliomedusa*).

Genus *Diandongia* Rong, 1974

Diandongia pista Rong, 1974

Specimens of this brachiopod are usually found preserved in three dimensions, but the valves are sometimes a little compressed and cracked.

The valves are ovoid in outline, broader than high, with the ventral valve slightly the larger of the two. The size of specimens normally ranges from 3×2 mm to 12×10 mm. The profile is biconvex, with the ventral valve more curved towards the umbo. Both valves display clear concentric growth lines, and some specimens show weak costae towards the anterior margin. A fine reticulate ornament is particularly well developed near the umbo. Internally on the ventral valve is a visceral platform, which does not extend to mid-length.

This species probably lived epifaunally, attached by a short pedicle.

Diandongia is a common brachiopod in the Chengjiang and Haikou areas, and is also found outside the Chengjiang biota.

Key References Rong 1974, Luo *et al.* 1999.

Figure 17.1 *Diandongia pista*. (a) Dorsal view of dorsal valve (RCCBYU 10291), ×10.0; Mafang. (b) Dorsal view of dorsal valve (RCCBYU 10289), ×6.7; Mafang. See also Fig. 17.2c, a ventral valve with a specimen of *Longtancunella chengjiangensis* pedically attached.

Genus *Longtancunella* Hou, Bergström, Wang, Feng & Chen, 1999

Longtancunella chengjiangensis Hou, Bergström, Wang, Feng & Chen, 1999

This brachiopod is very unusual, in that it mostly occurs as clusters conjoined by the pedicles, although isolated individuals occur. One specimen has been found attached by its pedicle to a brachiopod of the genus *Diandongia*. Preservation of both the shell and the pedicle is normally as a darkened flat impression, but specimens may show slight relief, particularly on the pedicle.

Both valves are of similar size, thin, slightly convex, and subcircular in outline. Maximum length is 21 mm and maximum width, which occurs about one-third of the distance from the posterior tip, is 19 mm, although some shells appear to be slightly broader than long. The original shell has been dissolved away, and neither ornament nor growth lines are apparent. Sparse, delicate setae are evident along the mantle margin, set about 1.0 mm apart and extending 1–2 mm beyond the edge of the shell. The pedicle is relatively short and robust, less than 10 mm long but 3 mm wide; annulation is very clear on the pedicles of some specimens.

Although most lingulid brachiopods lived buried in the sediment with only the anterior-most portion of the valves exposed, the apparently gregarious association of the specimens renders this unlikely for *Longtancunella*. The animal was probably an epifaunal suspension feeder, with the stout pedicle holding the animals clear of the muddy seabed.

Longtancunella is a rare component of the Chengjiang biota, but has been found in both the Chengjiang and Haikou areas.

Key Reference Hou *et al.* 1999.

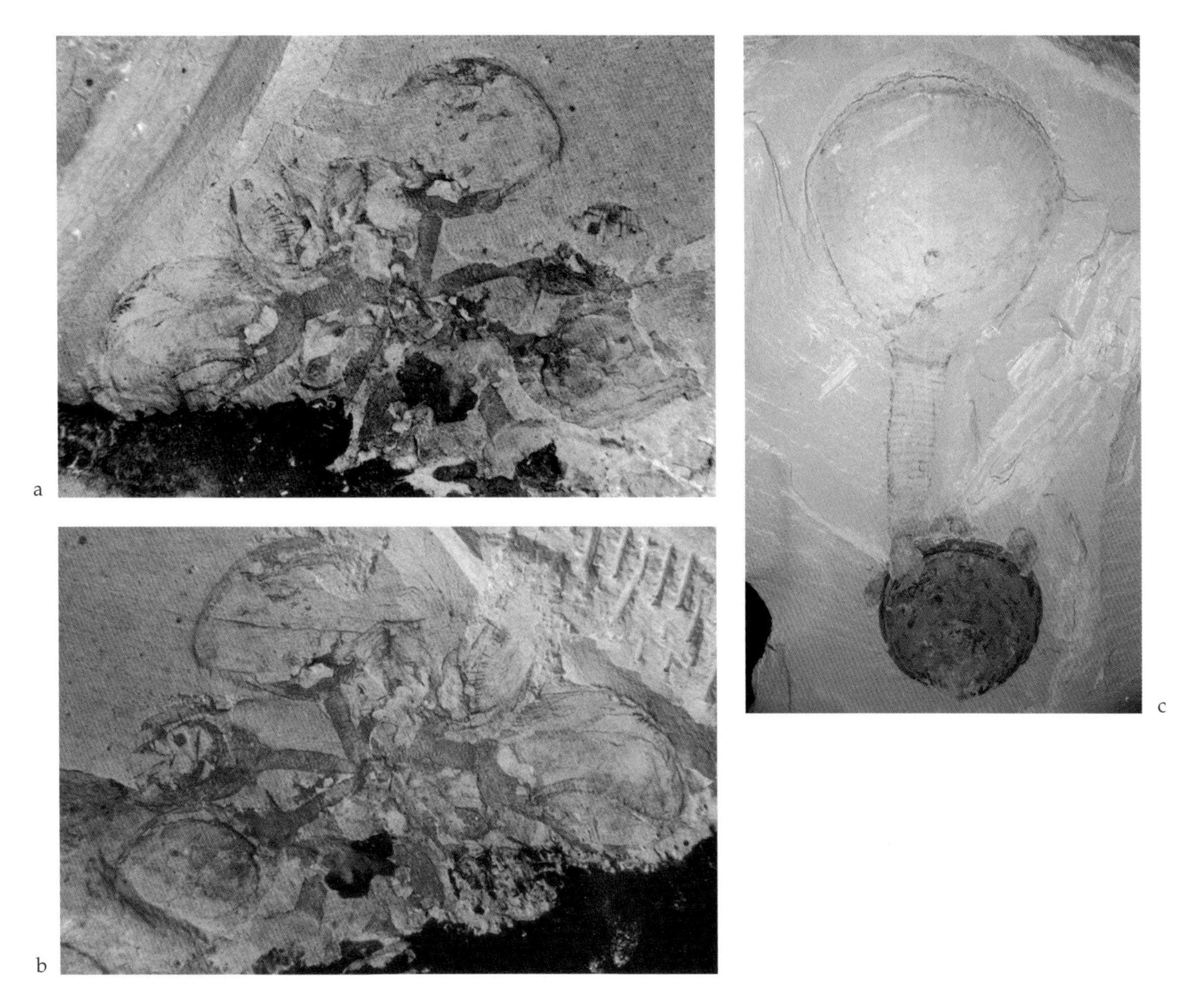

Figure 17.2 *Longtancunella chengjiangensis*. (a) Cluster of specimens with preserved pedicles, part (NIGPAS 11545a), × 2.4; Maotianshan. (b) Counterpart (NIGPAS 11545b), × 2.2. (c) Upper specimen, pedically attached to a ventral valve of *Diandongia pista* (RCCBYU 10292), × 1.7; Mafang.

Genus *Lingulellotreta* Koneva *in* Gorjansky & Koneva, 1983

Lingulellotreta malongensis (Rong, 1974)

This brachiopod is a common constituent of the Chengjiang biota. A remarkable feature is the spectacularly long pedicle, which is preserved in its entirety in many individuals. It is very rare to find the pedicle preserved in fossil brachiopods. Both the shell and the pedicle are preserved as flattened, slightly colored impressions.

The entire shell is 7–10 mm long and up to 5 mm wide. The shape is basically ovoid, but the ventral valve is conspicuously extended to a point at the posterior end, where a distinctly concave area (the pseudointerarea) is bisected by a narrow elongate pedicle foramen. Both valves are covered with discontinuous costae, which are of irregular width and become distinctly spinose at the anterior end. Both valves also show dense concentric growth lines, which may be grouped into bands about 1 mm wide. Internally, the dorsal valve shows fan-shaped muscle scars, divided into two parts by a strong median ridge; the interior of the ventral valve has not been described in Chengjiang specimens. The pedicle, when preserved, is some 50 mm long and 0.8–1.2 mm in diameter, becoming noticeably wrinkled or annulated towards the bulbous distal end.

Modern lingulid brachiopods of this type live in burrows, permanently attached by the contractile pedicle. *Lingulellotreta* may have adopted a comparable mode of life, but it may also have been epibenthic with the pedicle wholly or partly above the sediment (Holmer *et al.* 1997). It has also been suggested that this species was epiplanktonic, living attached to floating plants or animals. An epibenthic or epiplanktonic lifestyle could account for the very long pedicles and for the observation that representatives of this species are always preserved parallel to the bedding. Chen & Zhou (1997) have even suggested, imaginatively, that *Lingulellotreta* may have lived in burrows where the substrate was sandy, but adopted a pseudoplanktonic habit, attached, for example, to individuals of *Eldonia* and *Rotadiscus*, if the substrate was very soft.

L. malongensis has been found in both the Chengjiang and Haikou areas.

Key References Jin *et al.* 1993, Chen & Zhou 1997, Holmer *et al.* 1997.

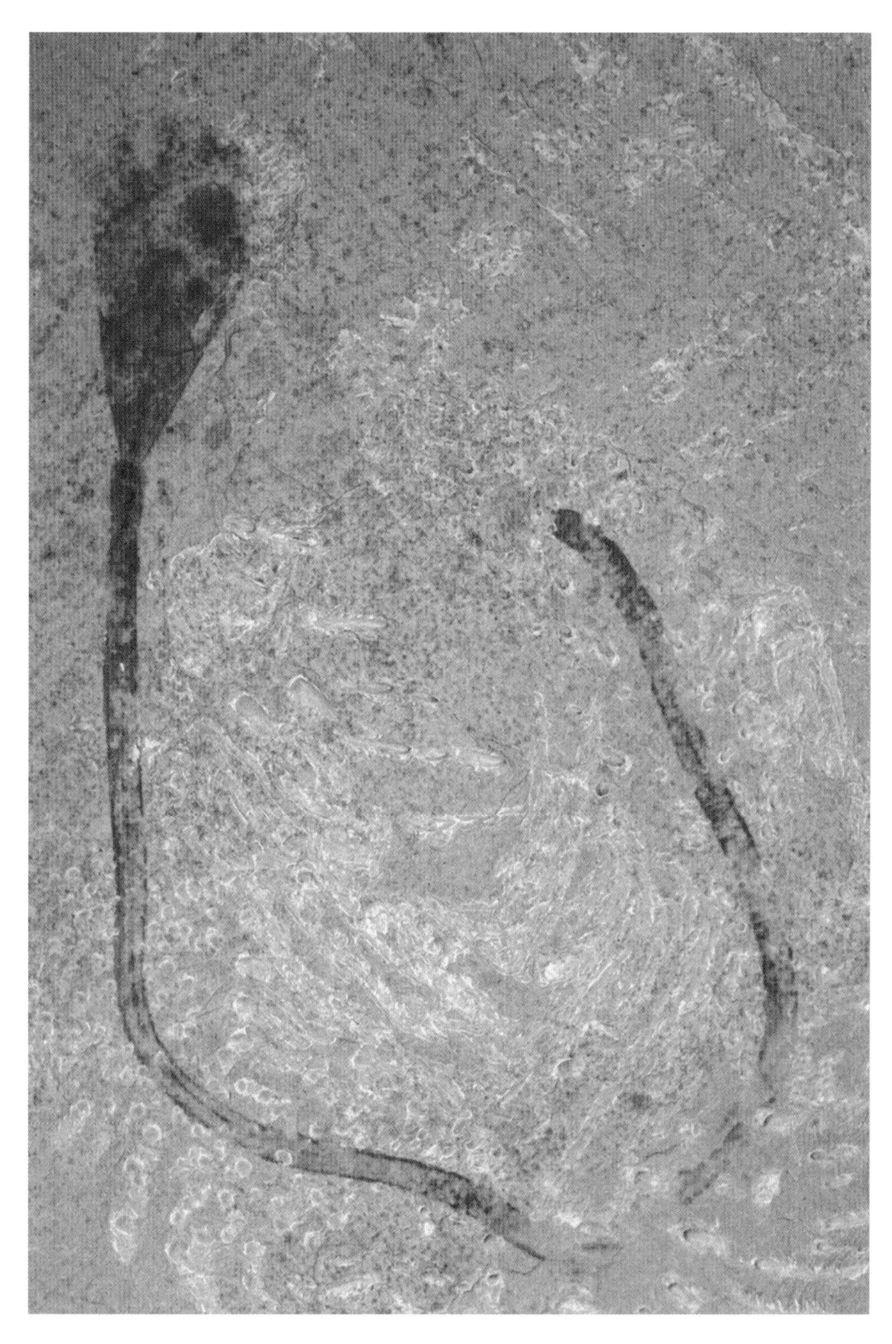

Figure 17.3 *Lingulellotreta malongensis*. Specimen with preserved pedicle (RCCBYU 10293), ×7.2; Mafang.

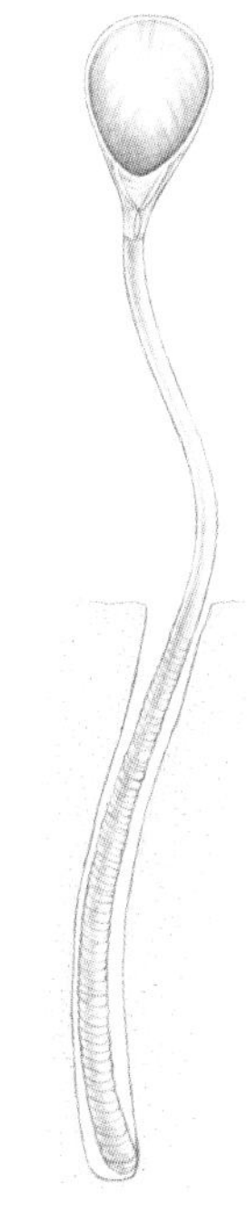

Figure 17.4 Reconstruction of *Lingulellotreta malongensis* in inferred life position.

Genus *Lingulella* Salter, 1866

Lingulella chengjiangensis Jin, Hou & Wang, 1993

This is a common fossil in the biota, and among the commonest of the brachiopods. As with *Lingulellotreta malongensis*, several specimens of this small brachiopod are characterized by the preservation of a remarkably long pedicle, although in some individuals the entire pedicle has not been preserved. Some specimens are currently being described in which the lophophore is also apparent (Zhang *et al.* in press). The valves and the pedicle are predominantly now flattened impressions, sometimes with a little relief at the margin of the shell.

The shells are up to 8.5 mm in length, with a triangular outline; the maximum width is close to the anterior margin. A distinct protegulum (the first-formed organic part of the shell) is evident in some specimens and is circular in outline. The thin valves are ornamented by fine, dense growth lines, which are grouped into well-defined bands. The ventral valve is extended posteriorly into a triangular, recurved pseudointerarea, divided by a deep pedicle groove. Muscle scars, forming a shield-shaped musculature area, have been described on the internal surface of the ventral valve. The pedicle reaches more than 60 mm in length and is 1.5 mm across; annulation is apparent along its length.

Lingulella is a very widespread and well-known Cambrian lingulid brachiopod; specimens occur in the Burgess Shale, although soft parts, including the pedicle, are not preserved there. This species may have adopted a comparable burrowing mode of life to modern lingulids but, as for *L. malongensis*, the long pedicle may attest to an epibenthic or epiplanktonic lifestyle. Chen & Zhou (1997) proposed that both of these species might have burrowed in sandy substrates, but that they adopted an epiplanktonic lifestyle, attached to floating organisms such as *Eldonia* and *Rotadiscus*, in conditions where the substrate became very soft and could not support burrows.

L. chengjiangensis has been recorded from both the Chengjiang and Haikou areas.

Key References

Jin *et al.* 1993, Chen & Zhou 1997.

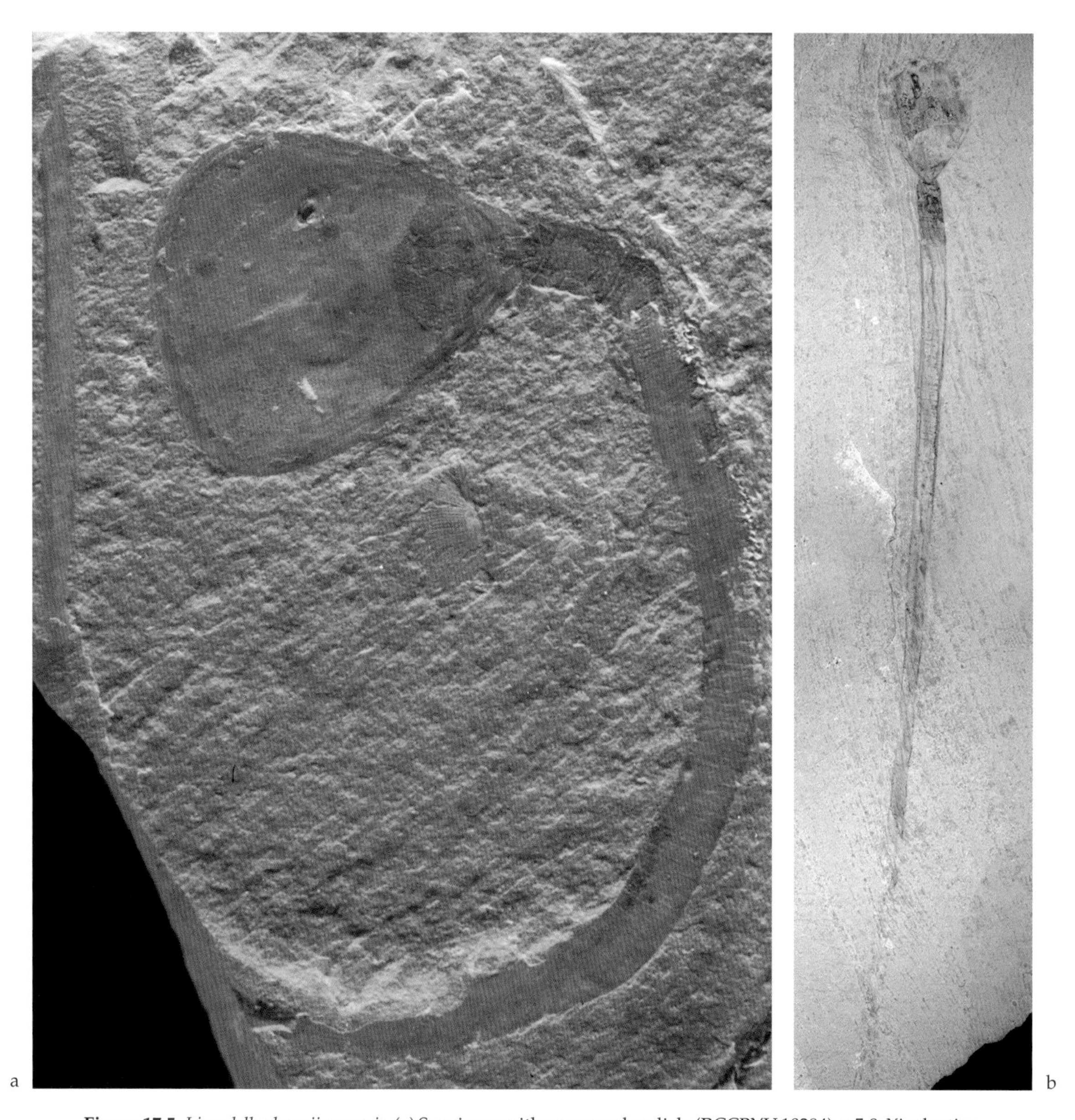

Figure 17.5 *Lingulella chengjiangensis*. (a) Specimen with preserved pedicle (RCCBYU 10294), ×5.8; Xiaolantian. (b) Specimen with preserved pedicle (RCCBYU 10295), ×2.5; Mafang.

Genus *Heliomedusa* Sun & Hou, 1987

Heliomedusa orienta Sun & Hou, 1987

This ovate brachiopod is one of the commonest animals in the Chengjiang fauna. The specimens are preserved as red-stained composite molds and casts, with internal features most strongly displayed. The specimens show considerable compaction with associated deformation, although some of the original convexity of the valves may remain. Preserved soft tissues reported from specimens of *Heliomedusa* include sensory setae, often apparent round the margin, the lophophore, mantle canals and traces of the nerves (Zhang *et al.* in press).

Both valves are convex, from 5 to 22 mm in length and generally slightly broader than long; the posterior margin is straight or very slightly curved. The shell is thin and appears to have been only weakly biomineralized; no trace of the original shell material remains. The ventral valve shows a beak-like umbo. The apex of the dorsal valve lies to the posterior of the center of the shell, while the ventral apex is at the posterior margin, where it is commonly seen to be flattened to form a concave, circular cicatrix, or attachment scar. Muscle scars are often evident, with an elevated main middle scar flanked by pairs of anterolateral and posterolateral scars. The surfaces of the valves show growth lines and indistinct fine radial ridges. Dense, short, fine setae extend along almost the entire margin of both valves; a few of these setae are distinctively coarser than their neighbors.

The ovate shape and radial patterning led the original authors to describe this animal as a jellyfish. Conway Morris & Robison (1988) indicated that it was a thin-shelled brachiopod, which was confirmed in a detailed study by Jin & Wang (1992), who considered that it was related to the craniopsids. Craniopsid brachiopods normally have a calcite shell, for which no direct evidence remains in *Heliomedusa*; notwithstanding this, Popov & Holmer (2000) placed the genus within the Craniopsida.

Heliomedusa lacks a pedicle, and the fine setae that occur round much of the margin suggest that this species did not burrow but lay on the surface of the sediment as an epifaunal suspension feeder (Chen & Zhou 1997). The cicatrix sometimes apparent at the ventral apex indicates that some or all individuals may have been attached to a hard substrate.

This species has a widespread distribution in eastern Yunnan Province.

Key References Sun & Hou 1987a, Jin & Wang 1992.

Figure 17.6 *Heliomedusa orienta*. (a) Dorsal view (RCCBYU 10297), × 5.4; Maotianshan. (b) Detail of anterior margin (RCCBYU 10296), × 10.4; Mafang. (c) Dorsal view (RCCBYU 10296), × 6.0; Mafang.

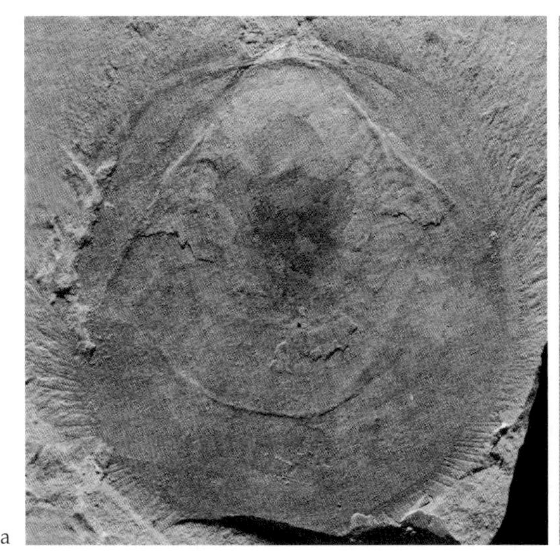
a

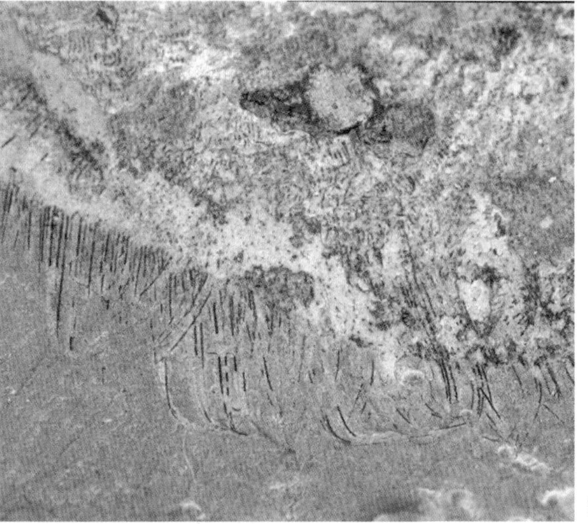
b

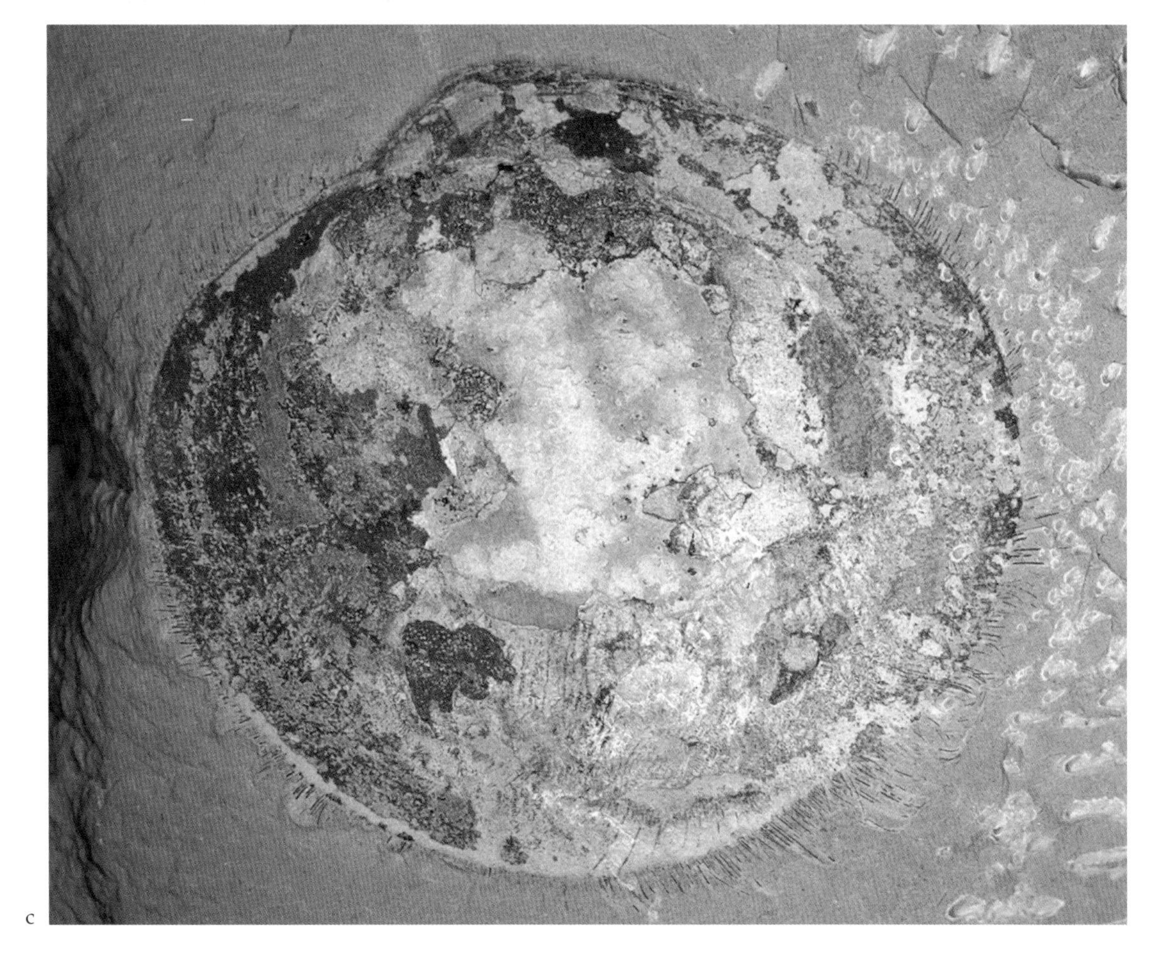
c

18 PHYLUM? VETULICOLIA

This group was recognized in 2001. *Vetulicola* was previously the only reasonably well-known genus, and it was usually regarded as an arthropod. With the idea that *Vetulicola*, *Banffia*, *Didazoon* and *Xidazoon* form an apparently natural group came the proposal that it be regarded as a distinct phylum, the Vetulicolia (Shu *et al.* 2001), placed among the primitive deuterostomes, perhaps with affinities to chordates. Lacalli (2002) suggested that vetulicolians might be urochordates. There is currently no agreement about the systematic position of the group.

The vetulicolians had a body divided into a bulky anterior part and a somewhat flattened tail. Both parts have a more or less clear segmentation, which is most obvious in species with a strongly cuticularized skin. In some species the cuticle of the anterior part forms paired dorsal and ventral plates, leaving between them a series of four to five lateral openings. The mouth opening is in the anterior end and appears to have been rounded. The anus is in the posterior end of the tail.

Vetulicolians have been suggested to be swimmers, feeding by means of ciliary filtering, or they may have been deposit feeders.

Genus *Vetulicola* Hou, 1987

Vetulicola cuneata Hou, 1987

This is a fairly common animal, usually found laterally compressed. The holotype specimen is 92 mm long and 32 mm high. The design is odd; at first sight it looks like a crustacean with a large carapace. However, the "carapace" is closed on the underside, and there are a series of four to five slits along the mid-line of each side; these have been interpreted by some authors to be gill pouches housing filamentous gills (Shu *et al.* 2001). Posteriorly on the dorsal side the "carapace" carries a triangular fin-like projection, and there are other, smaller projections evident on the posterior and anterior margins of some specimens. The abdomen has six segmental skeletal rings and a terminal telson. The telson and the last three segments each have a pair of horizontal flaps that give them an oval shape. This portion thereby appears to be a horizontal propulsive fin divided by three articulations. According to one report, however, the tail fin was asymmetrical and vertical (Shu *et al.* 2001).

No eyes or appendages are known. The narrow gut can be followed from the anterior part of the body, where it is straight, into the anterior part of the abdomen, where it spirals; it then becomes straight again and terminates at the posterior tip of the telson. *Vetulicola* was probably able to swim, and has been viewed by different scientists as a filter feeder, with gill pores open to the pharynx, or as a deposit feeder, with a primary sediment fill preserved in the gut.

Short specimens have been separated as a distinct species, *Vetulicola rectangulata*, but it may not be different. *Vetulicola* is known only from the Chengjiang fauna.

Key References Hou *et al*. 1999, Shu *et al*. 2001, Lacalli 2002.

Figure 18.1 *Vetulicola cuneata*. (a) Lateral view (NIGPAS 1100164), × 1.0; Maotianshan. (b) Lateral view (RCCBYU 10298), × 1.0; Dapotou.

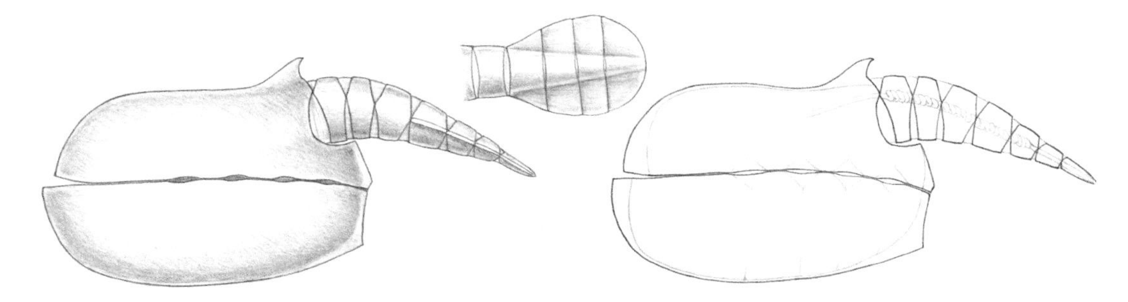

Figure 18.2 Reconstruction of *Vetulicola cuneata*, indicating in lateral views its external form and some internal features, and in dorsal view the oval-shaped posterior fin.

Genus *Banffia* Walcott, 1911

Banffia confusa Chen & Zhou, 1997

This is a fairly common species, which is usually poorly preserved as a thin film. The anterior part of the body is usually laterally compressed, but the "tail" normally appears to be dorsoventrally flattened, resulting in a twist in the middle of the specimen. Often only the posterior portion of the animal is preserved.

Banffia reaches a length of 55 mm. Like other vetulicolians, *Banffia* is divided into an anterior part and an abdomen or "tail". The anterior part is comparatively slender. Posteriorly, it extends into a low fin both dorsally and ventrally. Lateral slits have been described as extending longitudinally along most of the "carapace", but these are not clear. The tail is long, flat and broad. It is not divided into clear segments, but is fairly tightly wrinkled and appears to have been flexible. It is broad at the anterior end, but it is possible that it has a constriction at the attachment to the anterior part. Internally, the gut is represented by a dark, fine axial line that extends posteriorly to a terminal anus.

The mode of life of *Banffia* was presumably similar to that of *Vetulicola*. The animal appears to have had some swimming ability, propelled by the"tail", but there is no direct evidence from pharyngeal slits/pores or from sediment in the gut to demonstrate how it fed.

The species is restricted to the Chengjiang fauna, but the genus *Banffia* is known also from the Middle Cambrian Burgess Shale of British Columbia (Walcott 1911c).

Key References Chen & Zhou 1997, Hou *et al.* 1999, Shu *et al.* 2001.

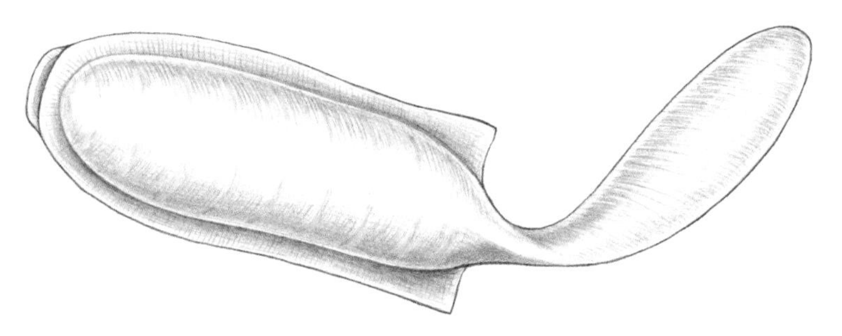

Figure 18.3 Reconstruction of *Banffia confusa*.

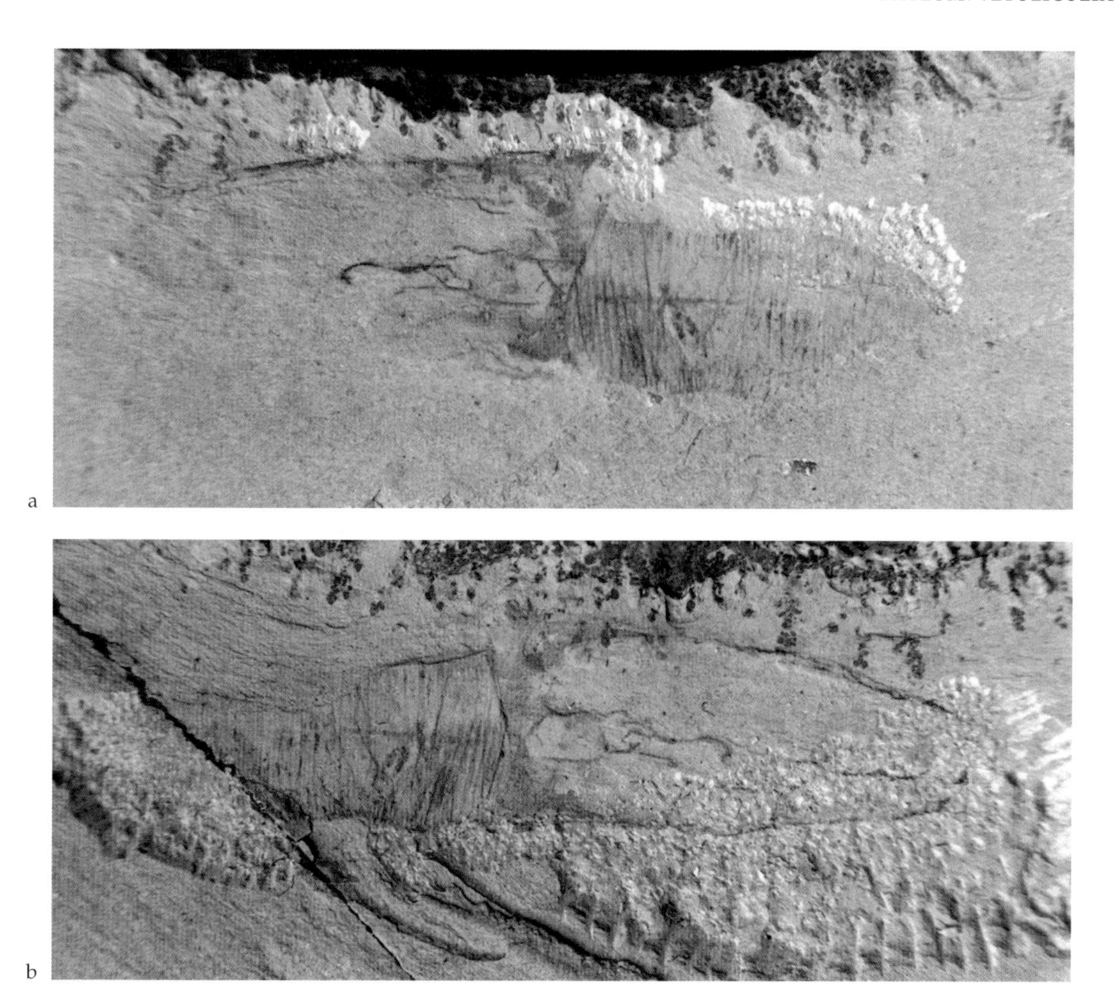

Figure 18.4 *Banffia confusa*. (a) A complete specimen, part (RCCBYU 10299), ×3.0; Xiaolantian. (b) Counterpart (RCCBYU 10300), ×3.0; Xiaolantian.

19 PHYLUM CHORDATA

The chordates are united by the possession of a notochord, a stiffening axial rod of tissue that becomes the backbone in the advanced vertebrates. Other characteristics include a dorsal neural tube above the notochord and ciliated pharyngeal slits. Some authorities divide the living chordates into three phyla: (i) the Urochordata (tunicates), in many of which the tadpole-like larva has a notochord which is lost in the sessile adults; (ii) the Cephalochordata, which have an unmineralized notochord, serial V-shaped muscles, and filter feed using the branchial basket; and (iii) the Vertebrata or Craniata, which are fundamentally characterized by the evolution of neural crest cells and epidermal placodes that give rise developmentally to a number of features of the skeleton and sensory organs.

Reports of chordates from the Chengjiang biota are relatively few. A single specimen of a possible urochordate was described by Shu *et al.* (2001), but its identity remains highly equivocal; it may be an incomplete specimen of the enigmatic genus *Phlogites*, described by Luo & Hu (*in* Luo *et al.* 1999). A probable cephalochordate, *Cathaymyrus diadexus*, has also been recorded (Shu *et al.* 1996), and two further specimens of a similar type were reported by Luo *et al.* (2001). The latter two fossils are possibly poorly preserved specimens of the worm *Cricocosmia*. Some authors have also regarded the enigmatic taxa *Yunnanozoon* and *Haikouella* to be chordates (see below). The most important specimens, however, are those representing the earliest known vertebrate, *Myllokunmingia*.

Genus *Myllokunmingia* Shu, Zhang & Han *in* Shu *et al.*, 1999

Myllokunmingia fengjiaoa Shu, Zhang & Han *in* Shu *et al.*, 1999

Myllokunmingia is one of the most celebrated fossils from the Chengjiang biota, as it is the earliest known vertebrate. Although not first described until 1999, more than 500 specimens have now been reported (Shu *et al.* 2003). The specimens are normally preserved in lateral aspect as flattened imprints, although there is a low relief that emphasizes several features of the animal.

All specimens illustrated in the literature display the trunk and/or head of the animal, but none preserves the tail. A series of V- or W-shaped muscle blocks is apparent along the trunk, and a complementary set of supposed gonads, at least 24 in number, is present towards the ventral margin. A distinct dorsal fin is present and there is a single, or possibly paired, smooth, narrower ventral fin. Traces of ray-like supports in the dorsal fin are inclined forwards toward the anterior, but may have different orientations on other parts of the fish. A trace of the gut has been identified on some specimens and there may be a subterminal anus. At the anteroventral end of the trunk there is a set of six, maybe seven, gill pouches, containing feathery gill filaments. A set of arched structures extends from behind the head for more than half the length of the animal (Hou *et al.* 2002a). The dorsal parts of these arches were interpreted as arcualia supporting a notochord (Shu *et al.* 2003), but their extent across the entire height of the body makes this interpretation unlikely. A possible pericardial cavity is evident above the ventral margin behind the gill pouches. The head is differentiated as an area of darker preservation; patches of differential coloring or relief in this region may represent preserved cartilages, and some specimens show eyes and other possible sensory structures.

The first two specimens described were assigned to separate genera, *Myllokunmingia* and *Haikouichthys*, but with the discovery of a third specimen, Hou *et al.* (2002a) concluded that only one species was represented by all the specimens. The presence of muscle blocks, a dorsal fin, filamentous gills and paired sensory structures in the head together demonstrate that the animal is a vertebrate, albeit without any phosphatic skeleton. Cladistic analysis shows it to be a primitive crown-group vertebrate, more derived than the hagfishes (Hou *et al.* 2002a).

Myllokunmingia was certainly able to swim, but it is not yet known how it fed.

This species is known only from the Haikou area of Yunnan Province.

Key References Shu *et al.* 1999, Hou *et al.* 2002a, Shu *et al.* 2003.

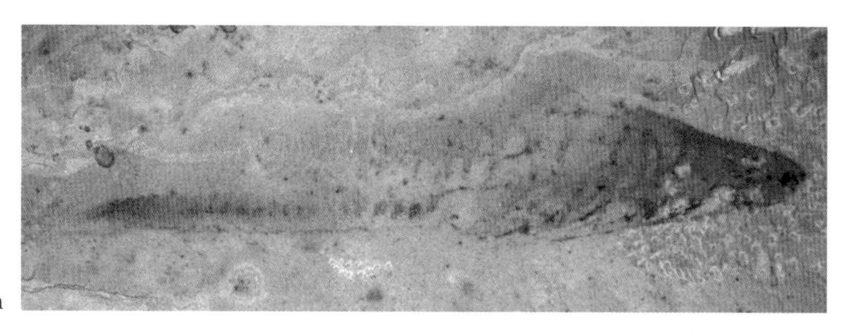

a

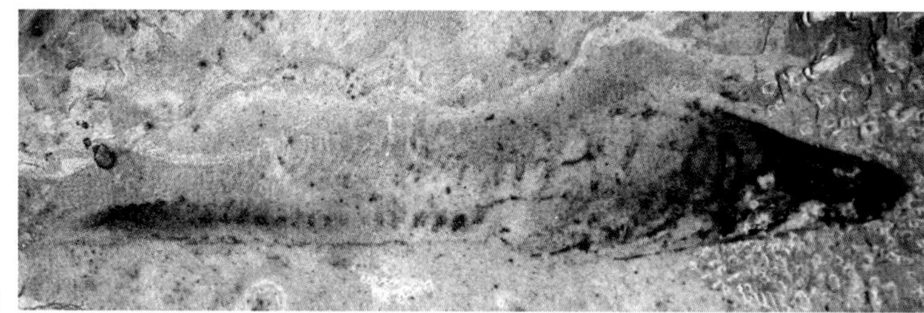

b

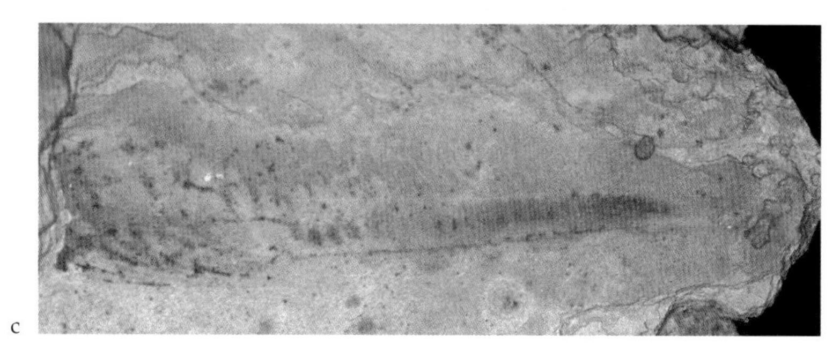

c

Figure 19.1 *Myllokunmingia fengjiaoa.* (a) Lateral view, part (RCCBYU 00195a), ×3.7; Ercaicun. (b) Lateral view, part (digitally enhanced), ×3.7. (c) Lateral view, counterpart (RCCBYU 00195b), ×4.0.

20 ENIGMATIC ANIMALS

Genus *Allonnia* Bengtson & Hou, 2001

Allonnia phrixothrix Bengtson & Hou, 2001

Several complete specimens of these chancelloriids are known from the Chengjiang biota. The specimens are 40 mm or more in length and comprise a sac-like body with a flexible skin (integument) covered with spiny sclerites. The fossils are compressed, but retain some relief and are commonly highlighted in the rock by a reddish coloring imparted by iron oxide.

The wall of each of the triradiate sclerites is thin and was probably originally calcareous; this wall encloses an internal cavity, now preserved as a clay infilling. Each ray is about 8 mm long and attached to the integument at its base; two of the rays are positioned close to the body surface with the third directed outwards. Between the sclerites the integument is folded, showing that it was flexible in life, and displays a conspicuous rhombic pattern that appears to be caused by small imbricating platelets, each about $30 \times 60\,\mu m$ in size. In places the platelets become elongated, perhaps forming small spinules that protruded from the surface.

Chancelloriids were initially interpreted as sponges (Walcott 1920), an opinion that has been followed and endorsed by many subsequent authors. Contrasting views have been put forward by other workers (Mehl 1996, Bengtson & Hou 2001, Janussen *et al.* 2002): Mehl, for example, considered that the sclerites were covered by an outer layer of tissue in life and that their mode of formation indicated an affinity with ascidians (tunicates). Bengtson & Hou (2001) disputed this assignment and provocatively pointed out similarities, and possible homologies, between the sclerites of chancelloriids and those of coeloscleritophorans, including the halkieriids. The debate is currently open.

Chancelloriid ecology is also unresolved. Bengtson & Hou (2001) regarded the sclerites of *A. phrixothrix* to be too scattered to have served a supportive function and suggested that their prime purpose was as protection against predators. The organisms were probably attached to the seabed at the narrower, basal end. They may have been suspension feeders, but there is no evidence in the fossils as to where nutrient-bearing water would have entered and exited the body cavity. Perhaps there was an apical opening or openings that cannot be observed on the laterally compressed fossils, or there may have been tiny openings in the integument that were involved in water circulation. There is no evidence of any prey-capture organs such as tentacles. Another possibility is that nutrition was obtained through a symbiotic relationship with algae or bacteria.

Chancelloriid sclerites are common fossils in Early Cambrian to early Late Cambrian strata, with complete specimens known from the Burgess Shale and the Chengjiang Lagerstätte. *A. phrixothrix* is known especially from localities near Haikou and is rarer in the Chengjiang area.

Key References Bengtson & Hou 2001, Janussen *et al.* 2002.

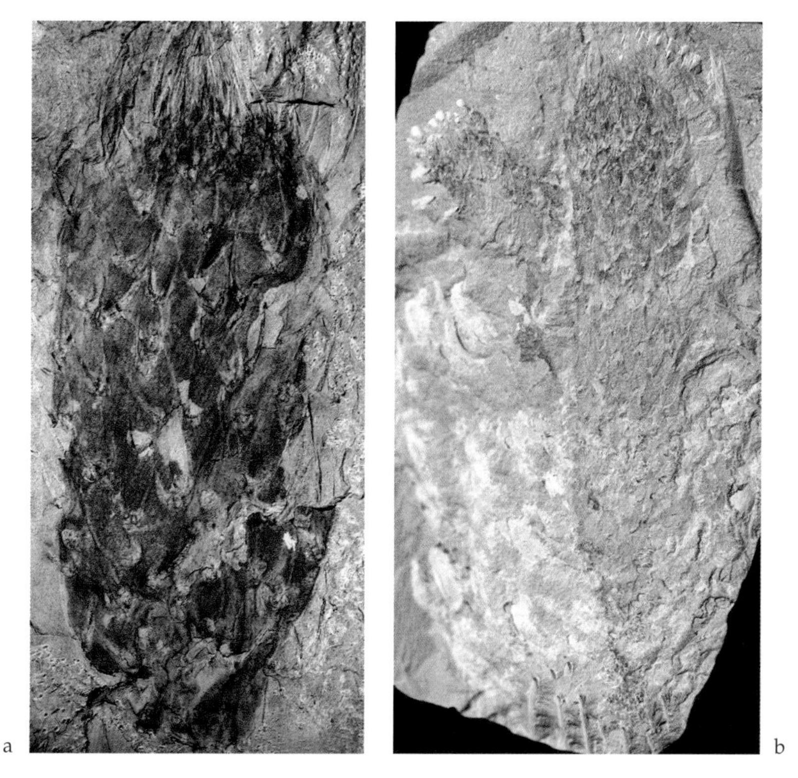

Figure 20.1 *Allonnia phrixothrix*. (a) A complete specimen (RCCBYU 10160), × 2.2; Xiaolantian (courtesy of Stefan Bengtson). (b) A complete specimen (RCCBYU 10301), × 1.8; Mafang.

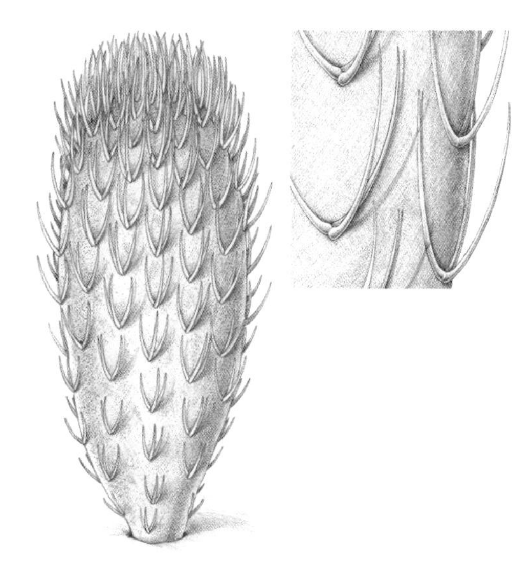

Figure 20.2 Reconstruction of *Allonnia phrixothrix* (after Bengtson & Hou 2001).

Genus *Batofasciculus* Hou, Bergström, Wang, Feng & Chen, 1999

Batofasciculus ramificans Hou, Bergström, Wang, Feng & Chen, 1999

Very few specimens are known of this strange, enigmatic fossil, which is more than 50 mm in length. Each specimen consists of a set of branches, at least 8–10 in number; each branch is independent, with a pointed base. The bases are set close together. From these bases the branches curve gently; they have smooth, flat surfaces, and each branch is ornamented on the convex margin by widely-spaced short spines that decrease a little in length away from the base. The spines normally show a very regular distribution, but in places the spacing becomes less consistent; they appear to have been rigid, perhaps hollow, and are sometimes preserved three-dimensionally with a blackened outer layer that may be the remnants of an organic wall. Each spine has a prominent axial ridge on the upper and lower surfaces. The tips of the branches are not preserved, but fine longitudinal grooves appear to be present away from the base, and one more completely preserved branch on the holotype becomes sinuous distally and lacks spines on this portion.

The nature of this organism is a mystery. It is probable that the whole structure was radially symmetrical in life, but this cannot be deduced with absolute certainty from the flattened impressions. The separate bases to the arms suggest that it might not be part of a larger organism. It may be a colonial structure. The presence of rigid organic walls would be consistent with an assignment to the graptolites (or other hemichordates), algae, or perhaps the hydrozoans; although the lack of apparent openings along the branches or in the spines perhaps rules out a hemichordate affinity.

This rare genus is only known from the Chengjiang biota.

Key Reference Hou *et al.* 1999.

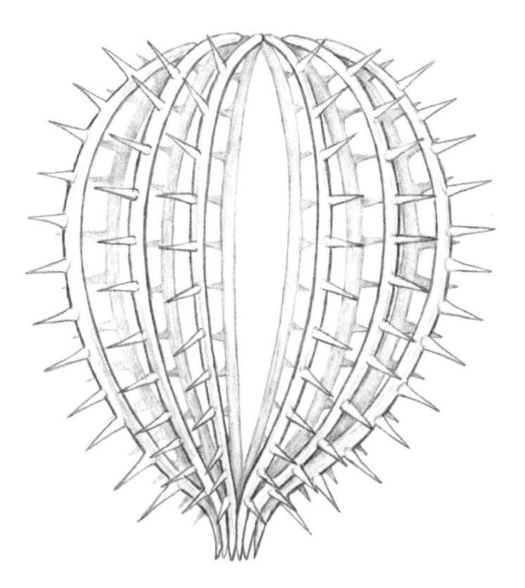

Figure 20.3 Reconstruction of *Batofasciculus ramificans*.

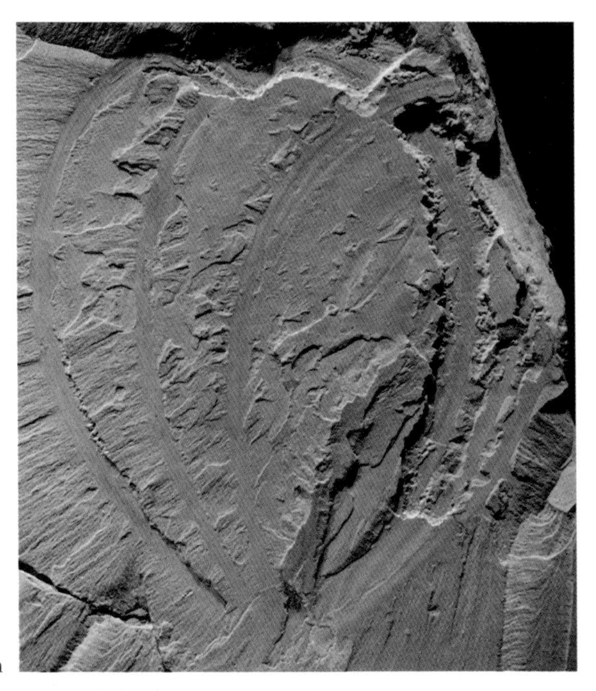

a

b

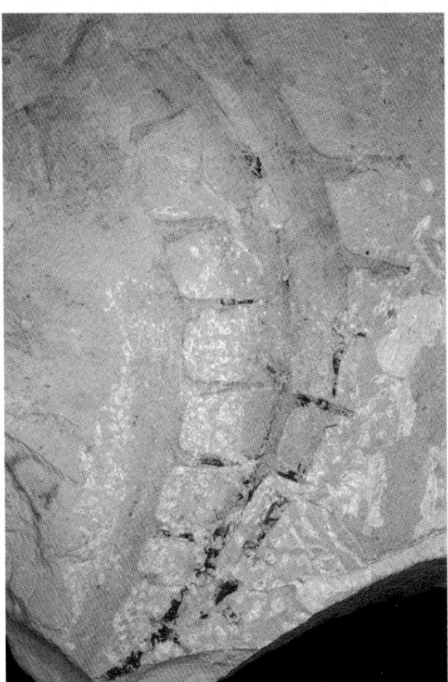

c

Figure 20.4 *Batofasciculus ramificans*. (a) A complete specimen, part (NIGPAS 115442a), × 1.6; Maotianshan. (b) A complete specimen, counterpart (NIGPAS 115442b), × 1.5. (c) Detail of spinose branches (RCCBYU 10302), × 2.6; Maotianshan.

Genus *Dinomischus* Conway Morris, 1977

Dinomischus venustus Chen, Hou & Lu, 1989

The fossils comprise a tall, slender stem, supporting a cone-shaped calyx with its upper margin fringed by a circlet of plate-like structures, termed bracts. The specimens can exceed 100 mm in height. Some specimens are preserved as impressions with low relief, but many are more three-dimensional.

The stem is tubular, terminating abruptly at the top, where the base of the calyx is connected; the base of the calyx is noticeably wider than the top of the stem. The calyx appears to be formed of two parts, one inside the other, together supporting about 18 rigid, plate-like bracts, about 12 mm in length, with wrinkled inner surfaces and outer surfaces divided into three smooth fields by longitudinal ridges. A single, tall, apparently tubular structure, 0.8 mm across, extends for some 44 mm above the bracts and may represent an excretory tube (Chen *et al.* 1989c) or a rudder for orientating the bract apparatus against the current (Chen & Zhou 1997). A large curved sac has been identified within the calyx.

Dinomischus was originally found in the Burgess Shale, where it is one of the rarest components of the biota. The preservation of the Chengjiang material is generally better, and the "excretory tube" has not been recognized in the specimens from Canada. However, the Burgess Shale specimens do reveal a structure interpreted as a U-shaped gut, a sac-like putative stomach, and also a bulbous swelling at the bottom of the stem that appears to be a holdfast that anchored the animal in the sediment (Conway Morris 1977c).

Dinomischus shows broad resemblances to a number of other stalked animals such as the echinoderms, but Conway Morris (1977c) considered that the fossil compares most closely to the Entoprocta. Dzik (1991), in contrast, considered that *Dinomischus* belonged with *Eldonia* and related animals in his new class Eldonioidea, for which he postulated affinities with the brachiopods and bryozoans.

Dinomischus probably lived on the seafloor with most of the stem clear of the sediment, holding the calyx above the seabed. The bracts were probably feeding devices, although their rigidity means that they were not true tentacles; it is possible that they were ciliated and swept suspended food into the mouth, which was situated on the upper surface of the conical calyx (Conway Morris 1977c). A specimen preserved with the bracts spread radially on the sediment surface indicates that they could have been spread in life to form a feeding bowl (Chen & Zhou 1997).

D. venustus has been reported only from Yunnan Province.

Key References Conway Morris 1977c, Chen *et al.* 1989, Chen & Zhou 1997, Hou *et al.* 1999.

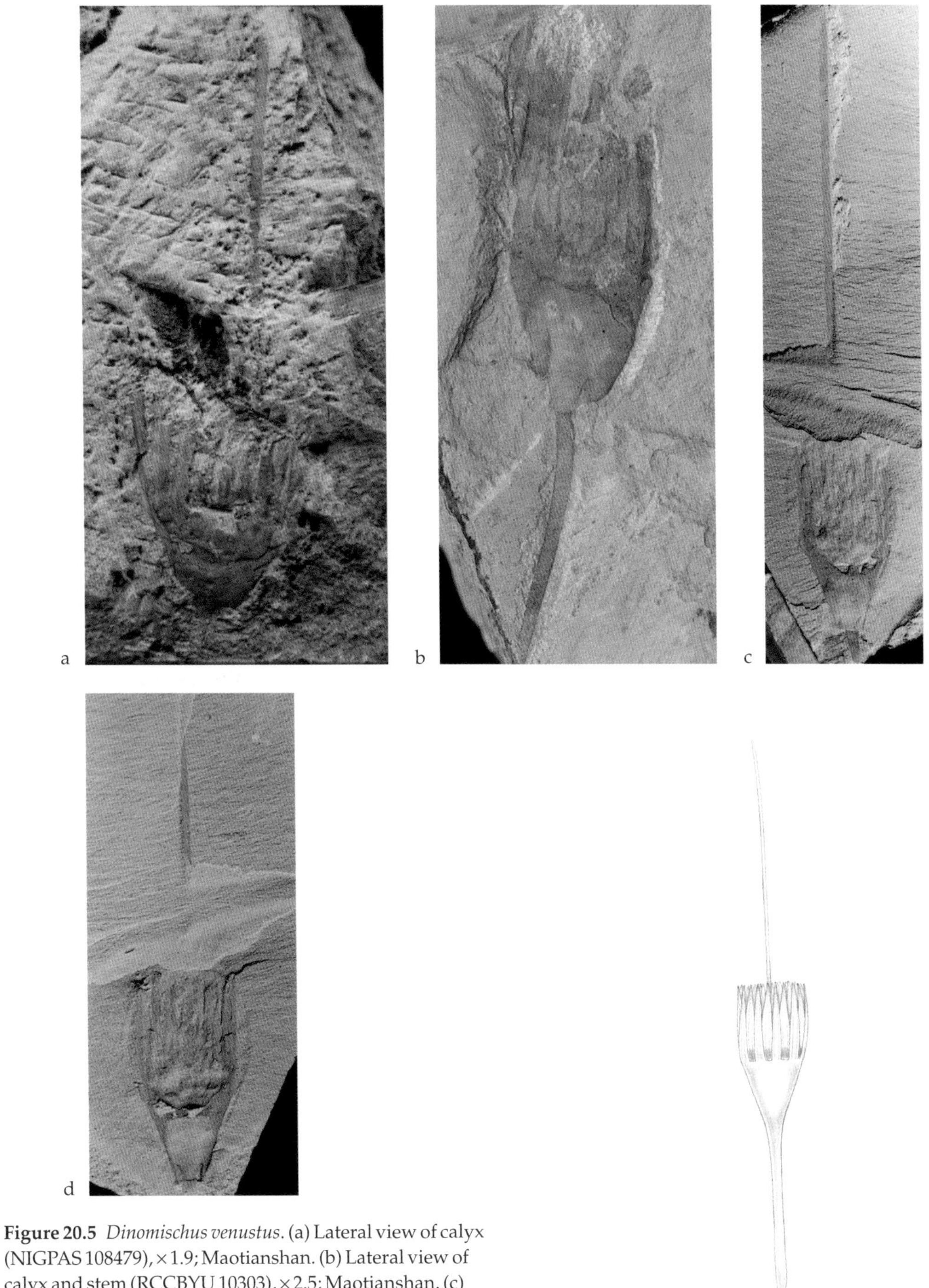

Figure 20.5 *Dinomischus venustus*. (a) Lateral view of calyx (NIGPAS 108479), × 1.9; Maotianshan. (b) Lateral view of calyx and stem (RCCBYU 10303), × 2.5; Maotianshan. (c) Lateral view, counterpart (NIGPAS 108478b), × 1.3. (d) Lateral view, part (NIGPAS 108478a), × 1.4; Maotianshan.

Figure 20.6 Reconstruction of *Dinomischus venustus*.

Genus *Eldonia* Walcott, 1911

Eldonia eumorpha (Sun & Hou, 1987)

This animal is one of the most frequent in the biota and often occurs in associations of several individuals. It is preserved as nearly flat discoidal impressions, but with a low relief defining the margin and picking out some internal structures.

The fossils are large, up to over 100 mm in diameter, with a circular outline; the presence of apparent concentric growth lines is evidence of marginal accretion. Distinct strands that radiate from the center to the margin have been interpreted as dorsal and ventral radial canals. There are about 80 dorsal canals; Chen & Zhou (1997) have suggested that these were fluid-filled and acted as a hydrostatic skeleton, but they may also be part of a water vascular system or internal support structures (Zhu *et al.* 2002). The ventral canals, about 40 in number, do not show three-dimensional preservation, and have been interpreted as mesenteries separating radiating lobes (Zhu *et al.* 2002). A U-shaped or dextrally coiled structure, sometimes darkened in color, surrounds the center of the disc at about two-thirds of the distance from the margin. This is considered to be the gut, with the oral and anal ends situated close to each other. A pair of tentacular structures surrounds the mouth in several specimens, each set ramifying distally.

The animal was originally described as a medusoid, under the name of *Stellostomites eumorphus* (Sun & Hou, 1987), but was subsequently considered to be a species of *Eldonia* (Conway Morris & Robison 1988). Walcott (1911b) thought that *Eldonia* was a pelagic holothurian (sea cucumber), a contention supported by several subsequent authors, although others have assigned it to the cnidarians. Dzik (1991) noted that there was no evidence of pentameral symmetry or of a mesodermal calcitic skeleton in *Eldonia*, making a holothurian affinity difficult to support. He suggested more radically that *Eldonia* and similar animals were characterized by the possession of a lophophore and thereby related to brachiopods and bryozoans, and proposed a new Class Eldonoidea to accommodate *Eldonia* and its relatives. Chen & Zhou (1997) further speculated that *Eldonia* and the related *Rotadiscus* evolved through heterochronic retention of the larval stage of a *Facivermis*-like ancestor, but there is no evidence to support this contention.

As well as being common in the Chengjiang biota, *Eldonia* is a well-known component of the fauna of the Burgess Shale, where it is represented by *Eldonia ludwigi* Walcott, 1911. Zhu *et al.* (2002) considered that there were sufficient differences between the Chengjiang and Burgess Shale specimens to warrant generic separation, but we have retained the name *Eldonia* for the Chengjiang material to emphasize the similarities.

The ecology of *Eldonia* is controversial. Its medusoid shape perhaps suggests that it was pelagic. Chen & Zhou (1997) noted that specimens of *Eldonia* sometimes occurred in close association with the lobopodians *Microdictyon* or *Paucipodia*, and concluded that the latter two were pseudopelagic riders on their pelagic host. Dzik *et al.* (1997), however, argued for a very different mode of life, considering the eldonioids to be benthic epifaunal animals, lying passively on the mud surface.

E. eumorpha has been recorded only from the Chengjiang biota.

Key References Sun & Hou 1987a, Chen *et al.* 1995e, Chen & Zhou 1997, Zhu *et al.* 2002.

Figure 20.7 *Eldonia eumorpha*. (a) A single specimen (RCCBYU 10304), × 1.1; Maotianshan. (b) A group of three specimens (RCCBYU 10305), × 0.6; Maotianshan.

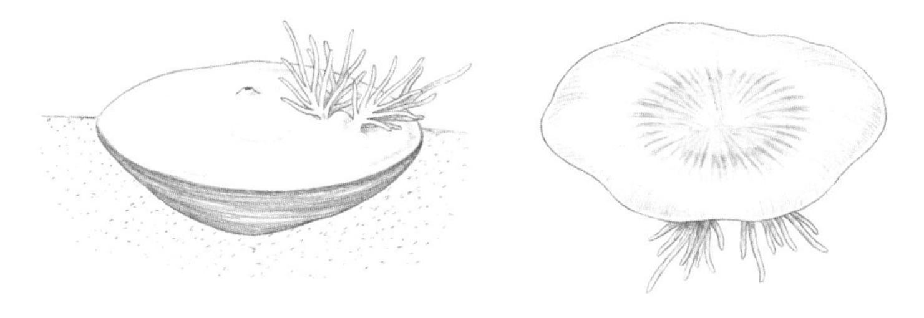

Figure 20.8 Reconstructions of *Eldonia eumorpha* as a sessile benthic organism (after Dzik *et al*. 1997) and as a pelagic organism (after Chen & Zhou 1997).

Genus *Facivermis* Hou & Chen, 1989

Facivermis yunnanicus Hou & Chen, 1989

This is a bilaterally symmetrical, worm-like animal, comprising a long, straight segmented trunk supporting a set of anterior tentacles. The holotype is 24 mm long in its preserved portion, but specimens occur up to 70 mm in length.

The trunk appears to have been originally cylindrical, parallel-sided for much of its length, but tapering anteriorly towards the tentacles. It shows clear annulations or segments, spaced with a density of 50–60 per 10 mm; there is no evidence of setae or parapodia extending from the trunk. A straight gut running medially in the trunk is preserved as a line of raised relief or as a black organic film. One specimen is reported as showing a mineralized tube around the posterior part of the trunk. At the anterior end, the front five segments bear symmetrically disposed tentacles, five on each side of the animal. Each tentacle is annulated, with each of the annuli bearing a long, fine spine with a thickened base. A small subtriangular area in front of the tentacles has been interpreted as a proboscis.

The biological affinities that have been proposed for *Facivermis* are fascinatingly diverse. Initially it was suggested that it might be related to the annelids, with attention drawn to the fact that the number of tentacles in the head region is identical to that of the extant nereid polychaetes. A relationship to lobopodians was subsequently suggested (Delle Cave & Simonetta 1991, Hou *et al.* 1991), but Hou & Bergström (1995) pointed out that the presence of spines on the tentacles of a new specimen refuted this. In a more radical interpretation, Chen & Zhou (1997) proposed that the oral tentacles are homologous to the lophophore and designated *Facivermis* as a member of the "Superphylum Lophophorata". A contrasting idea envisages a possible distant relationship to the pentastomids, a rather enigmatic group of parasitic arthropods (Delle Cave *et al.* 1998).

Facivermis is generally considered to have been a burrower (Hou & Chen 1989). A reconstruction (Chen *et al.* 1996) shows the trunk within the sediment and the tentacles collecting food from above the burrow opening. This interpretation is supported by the fact that specimens are commonly preserved slightly oblique to the bedding planes.

This rare genus is only known from a single species, from the Chengjiang biota.

Key References Hou & Chen 1989a, Chen & Zhou 1997.

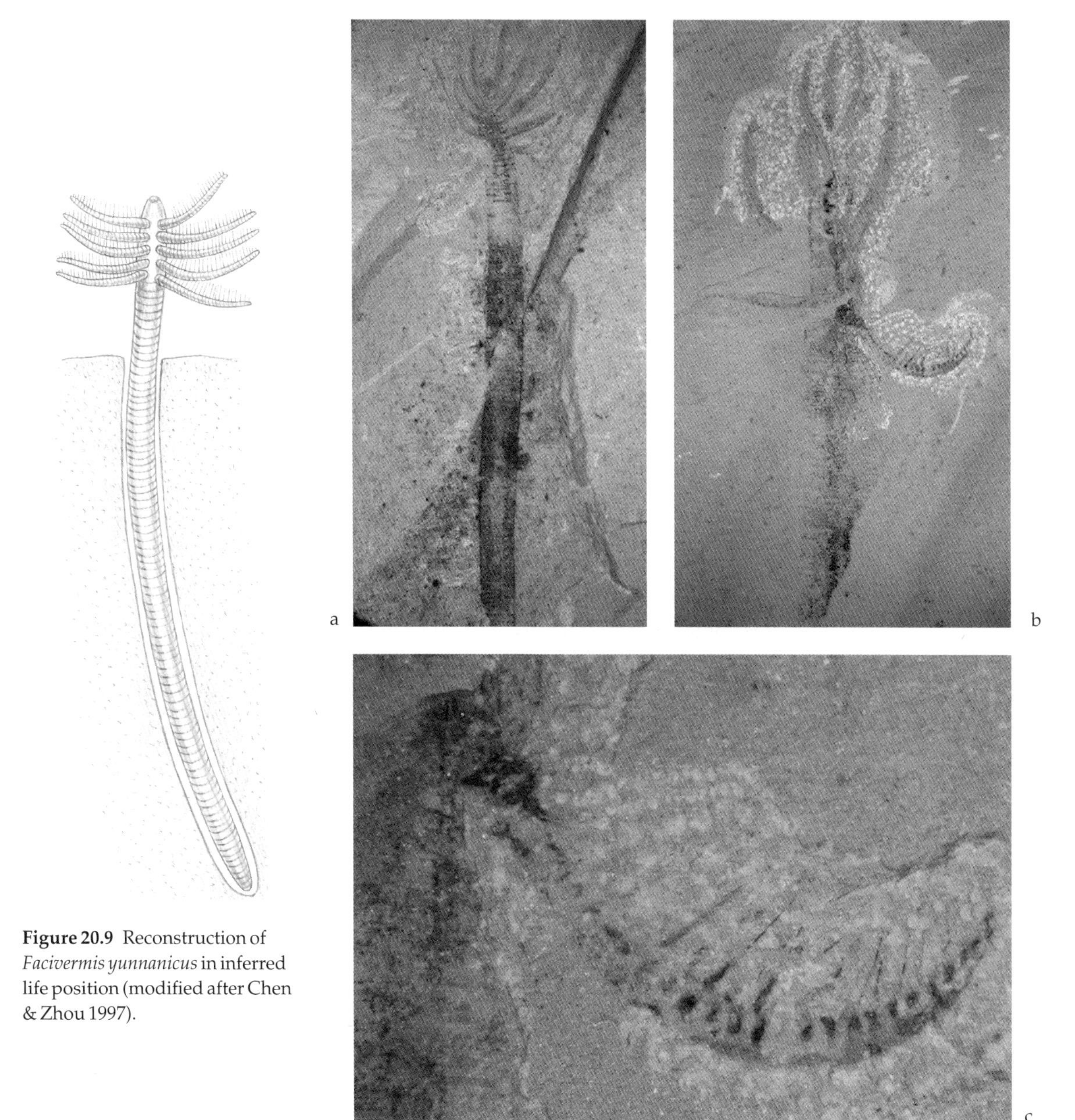

Figure 20.9 Reconstruction of *Facivermis yunnanicus* in inferred life position (modified after Chen & Zhou 1997).

Figure 20.10 *Facivermis yunnanicus*. (a) Anterior part of specimen (NIGPAS 108720), ×3.0; Maotianshan. (b) Anterior part of specimen (RCCBYU 10306), ×2.7; Maotianshan. (c) Detail of a single tentacle (RCCBYU 10306), ×11.7.

Genus *Jiucunia* Hou, Bergström, Wang, Feng & Chen, 1999

Jiucunia petalina Hou, Bergström, Wang, Feng & Chen, 1999

This is a rare, petal-shaped fossil preserved in low relief. Only the two associated specimens shown have been reported.

The complete specimen is 20 mm in length and 8 mm in maximum width and appears to be attached to the second specimen. At the attached end, the fossil narrows to a rounded point; the other end is broader, narrowing gently to a straight termination. The organism is divided longitudinally by three wide ridges, one medial and one on each side; together these divide the fossil into four areas of equal width. Narrower ridges are visible between the wide ridges, and the whole surface is covered by a fine, rather irregular, net-like ornament. Along both sides sets of dense setae or spicules are apparent, producing a serrated effect at the edge; one margin also shows close-packed tiny pores. The unattached surface also displays short spicules extending upwards beyond the margin.

The shape of the fossils suggests that they are flattened cups. Hou *et al.* (1999) considered that the net-like structures and spines might indicate a relationship to sponges, and if this is not a sponge it is likely that it is a colonial animal.

This rare animal is only known from the Chengjiang biota.

Key Reference Hou *et al.* 1999.

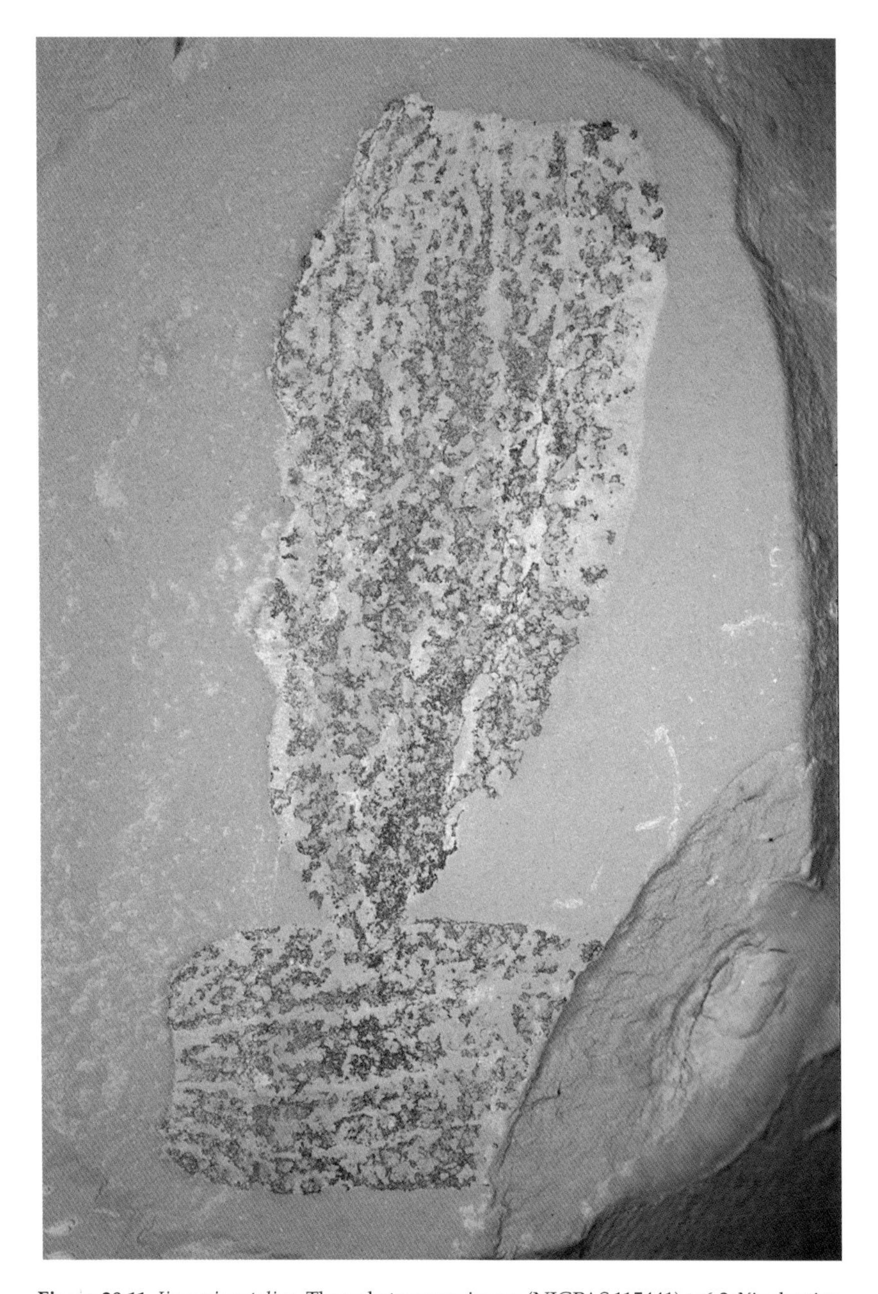

Figure 20.11 *Jiucunia petalina*. The only two specimens (NIGPAS 115441), ×6.2; Xiaolantian.

Genus *Maanshania* Hou, Bergström, Wang, Feng & Chen, 1999

Maanshania crusticeps Hou, Bergström, Wang, Feng & Chen, 1999

This is a fairly large worm-like animal, known only from a single specimen, 50 mm in length and up to 5 mm in width.

At one end is a subrectangular structure, preserved with strong relief and ornamented with numerous fine, longitudinal striae. This structure is 5 mm long and narrows from 5 mm to 3 mm wide towards the tip; it is most strongly convex where it is broadest. This feature was described by the original authors as a mineralized head, with papillae arranged in a transverse row on the surface. The trunk is annulated with 3–4 annuli per millimeter, and papillae are identified between the annuli, arranged in transverse rows. There is a wide straight gut trace medially, showing up as patches of darker coloration. The end opposite the smooth "head" is not preserved and the fossil fades distally in that direction.

Hou *et al.* (1999) noted that *Maanshania* is similar to paleoscolecid worms in the presence of transverse rows of papillae. They also noted similarities to lobopodians, but the fossil does not show any legs.

This genus is only known from *M. crusticeps* of the Chengjiang biota.

Key Reference Hou *et al.* 1999.

Figure 20.12 *Maanshania crusticeps*. The only known specimen (NIGPAS 115440), ×3.2; Ma'anshan.

Genus *Parvulonoda* Rigby & Hou, 1995

Parvulonoda dubia Rigby & Hou, 1995

This enigmatic species is known from only a few specimens, preserved either as flattened impressions or in bas-relief. Some node-like surface structures appear to be preserved as carbon films or organic outlines.

P. dubia is a small, strip-like, slightly curved and thin-walled conico-cylindrical fossil. Specimens are up to 16 mm long and 5 mm wide. Their surface contains several rows of 3–5 small hook-like nodes or low asymmetrical spines with circular bases. These structures, which range up to 1.5 mm across, are arranged in moderately regular, staggered or *en échelon* fashion and seem to be more prominently developed in the narrower (proximal?) part of the fossils. Conclusive spicule impressions have not been detected.

The nodose surface of *P. dubia* recalls that of the Burgess Shale sponge *Sentinellia draco* Walcott, 1920, which differs in its much larger overall size, larger nodes and in preserving convincing evidence of spicules (Rigby & Hou 1995). Without firm evidence of spicules the notion that *P. dubia* is possibly a sponge remains uncertain. From its general form and some aspects of its surface morphology the species has also been likened to the narrow, frond-like Burgess Shale alga *Margaretia dorus* Walcott, 1931 (Rigby & Hou 1995).

This rare species is known only from the Chengjiang biota.

Key Reference Rigby & Hou 1995.

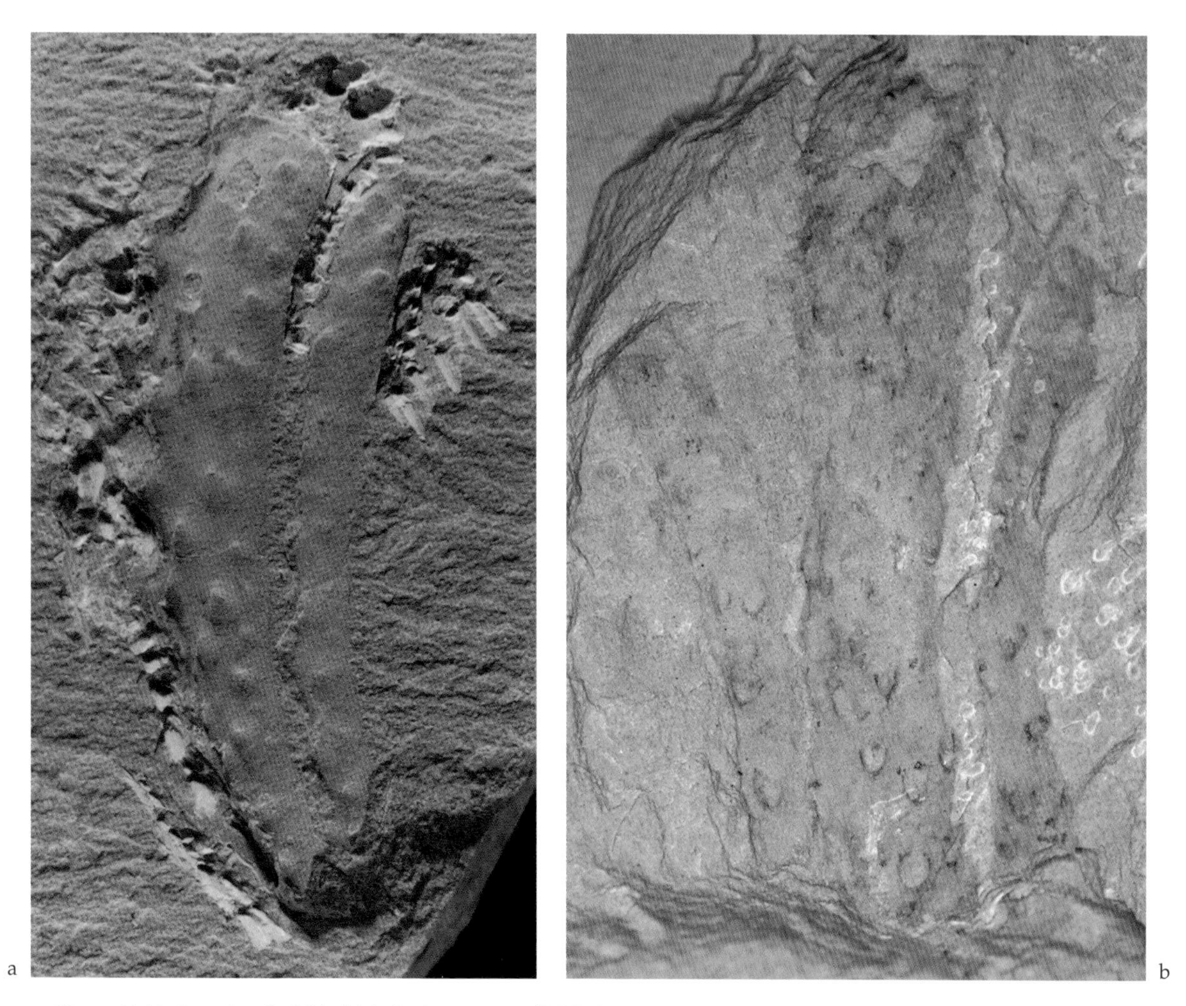

Figure 20.13 *Parvulonoda dubia*. (a) A single specimen (RCCBYU 10312), ×7.7; Maotianshan. (b) Possibly three specimens (RCCBYU 10307), ×6.6; Mafang.

Genus *Rotadiscus* Sun & Hou, 1987

Rotadiscus grandis Sun & Hou, 1987

Several tens of specimens are known of this large discoidal animal, up to 150 mm in diameter, preserved as flat impressions with little relief.

Rotadiscus has been described as having a sclerotized dorsal surface, covered with numerous fine radial lines and closely-spaced concentric rings; there are spots arranged radially between the radial lines. The ventral surface is separated from the dorsal disc and has a number (88 according to Zhu *et al.* 2002) of paired radial structures. A coiled sac within the disc appears to be the gut. A ramifying tentacular apparatus is associated with several specimens and has been reconstructed as hanging down below the radial umbrella (Chen *et al.* 1996). Some specimens in the Chengjiang fauna have been found with juvenile individuals of the lingulate brachiopod *Lingulellotreta malongensis* attached to them (Chen & Zhou 1997).

Rotadiscus was first described as a medusoid, but this was disputed by Dzik (1991) who included the genus with *Eldonia* in his new Class Eldonioidea, which he considered to be a group of lophophore-bearing animals. Closely similar specimens have been reported outside the Chengjiang biota, from the Middle Cambrian Kaili fauna of Guizhou Province. Some of these specimens of *Pararotadiscus guizhouensis* (Zhao & Zhu, 1994) are encrusted by shelled epizoans of unknown affinities (Dzik *et al.* 1997).

If *Rotadiscus* were a medusoid, it would be regarded as planktonic, but the reports of specimens supporting epibionts while alive led Dzik and his co-authors (1997) to propose that they were, in fact, sedentary animals that lay passively on the seafloor with the convex side of the disc facing downwards. In contrast, Chen & Zhou (1997) regarded the genus as pelagic and considered the encrusting epibionts to have adopted a pseudopelagic life strategy. This disparity in interpretations is the same as that outlined for *Eldonia eumorpha*.

Key References Sun & Hou 1987a, Chen & Zhou 1997, Dzik *et al.* 1997, Zhu *et al.* 2002.

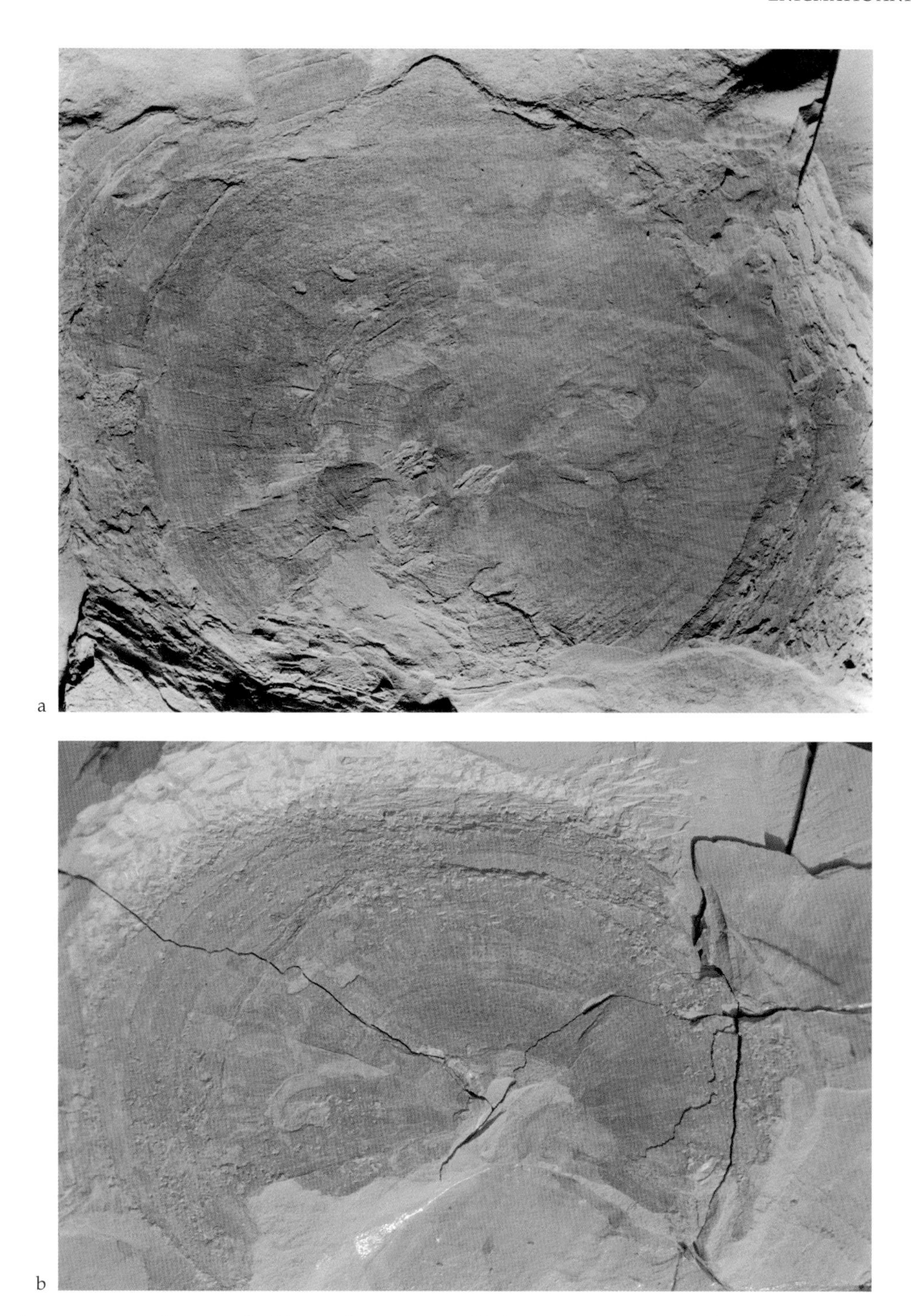

Figure 20.14 *Rotadiscus grandis*. (a) A single specimen (RCCBYU 10308), ×1.3; Maotianshan. (b) A single specimen (RCCBYU 10309), ×1.1; Xiaolantian.

Genus *Yunnanozoon* Hou, Ramsköld & Bergström, 1991

Yunnanozoon lividum Hou, Ramsköld & Bergström, 1991

Y. lividum is known from many hundreds of individuals, and slabs have been found covered with the remains of numerous specimens. It is an elongate, superficially worm-like animal, 25–40 mm in length. Nearly all the known specimens are preserved in lateral aspect, suggesting that the animal was laterally flattened in life. Preservation is as a blackened film, presumably carbonaceous.

The dorsal part of the body is divided into 22–24 subrectangular segments; the anterior four have been described as overlapping each other (Hou *et al.* 1999), but other authors have interpreted the segments as muscle blocks (myomeres) separated by myosepta (e.g. Chen *et al.* 1995a). Beneath the segments is an apparent gut, which in some specimens shows sedimentary infilling preserving a cryptically spiral shape, perhaps indicating the presence of a spiral valve. A row of small, paired subcircular structures along the ventral part of the animal has been interpreted as representing a set of gonads. In the anterior half there is a series of six or seven pairs of semicircular structures, often well preserved, dark in color and each composed of about 20–25 small discs that support short spines. This has been interpreted as a suite of branchial arches, perhaps external (Shu *et al.* 2003), but their interpretation as gills is equivocal (Bergström *et al.* 1998). A pair of cone-like structures has been reported in some specimens between the sixth and seventh "arches". At the anterior end there is an apparent head; this appears bilobed in some specimens, while others show an anterior opening, presumably a mouth, flanked by a lip-like structure.

The animal was originally designated as a worm-like animal of unknown affinities, but was subsequently considered to be a chordate by a number of authors, who interpreted the segmental structures as muscle blocks, although they are not V- or W-shaped and do not appear to flank a notochord. Other authors preferred an assignment to the enteropneusts, a group of hemichordates, with the "segments" attributed to a dorsal fin. Budd & Jensen (2000) have suggested that *Yunnanozoon* may be a stem-group taxon within the deuterostomes (the clade that includes echinoderms, hemichordates, and chordates); Conway Morris (2000) and Shu *et al.* (2003) offered a similar interpretation. Careful examination is still required before an accurate anatomical reconstruction can be attempted.

The ecology of *Yunnanozoon* is as enigmatic as its anatomy. Some specimens apparently have guts filled with mud, suggesting that the animal was a deposit feeder.

Yunnanozoon is unknown outside the Chengjiang biota. Specimens from the Chengjiang fauna assigned to *Haikouella* by Chen *et al.* (1999) and Shu *et al.* (2003) are closely similar to *Yunnanozoon* and may well represent the same genus, with the differences being taphonomic rather than anatomical. A separate status for *Haikouella*, however, has been recently advocated by Holland & Chen (2001), who placed this genus as a stem-group vertebrate.

Key References Hou *et al.* 1991, Chen *et al.* 1995a, Dzik 1995, Shu *et al.* 1996, Bergström 1997, Chen & Li 1997, Budd & Jenson 2000, Dewell 2000, Shu *et al.* 2003.

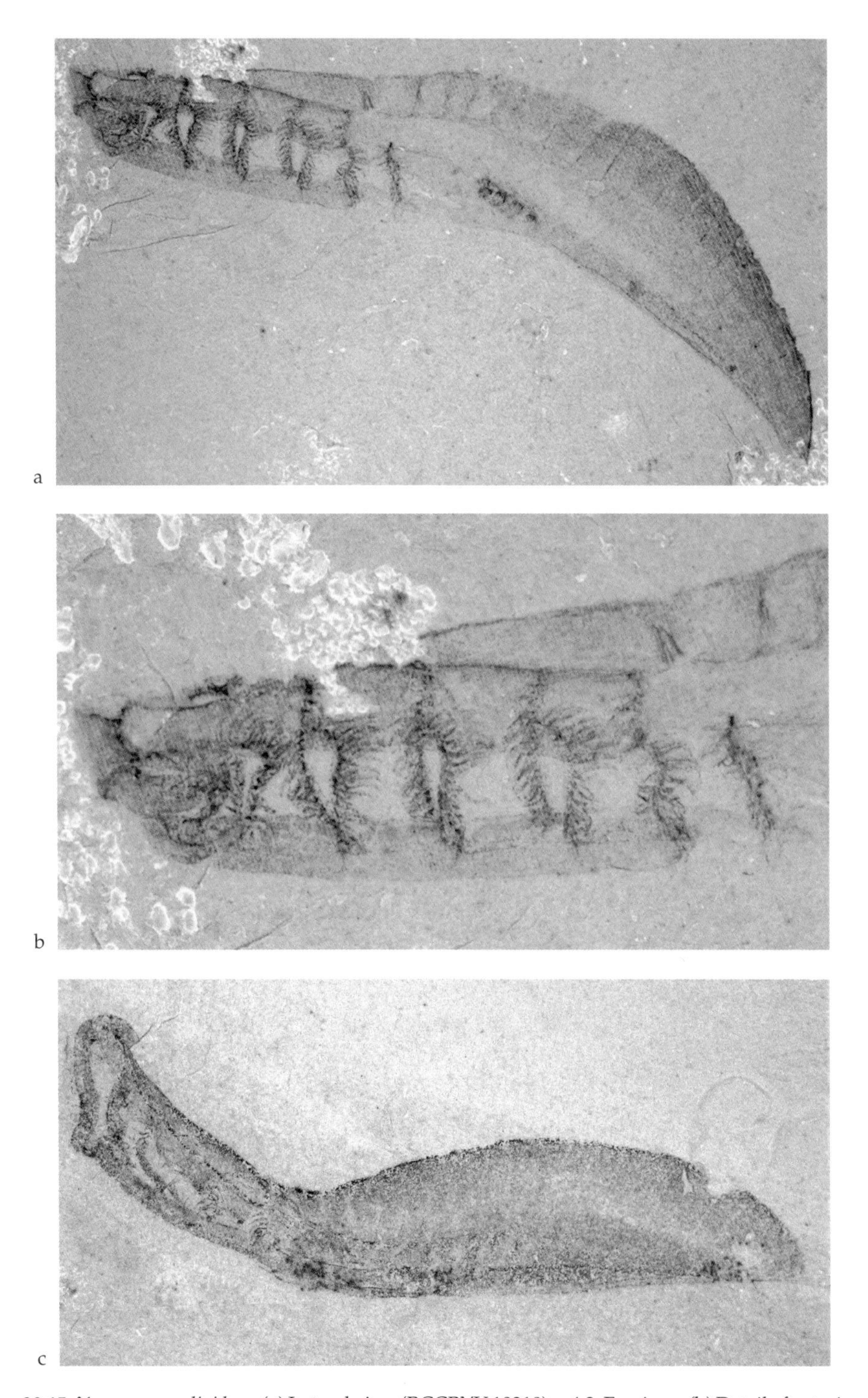

Figure 20.15 *Yunnanozoon lividum*. (a) Lateral view (RCCBYU 10310), ×4.2; Ercaicun. (b) Detail of anterior portion (RCCBYU 10310), ×4.2. (c) Lateral view (RCCBYU 10311), ×8.5; Ercaicun.

21 SPECIES RECORDED FROM THE CHENGJIANG BIOTA

Chengjiang fossils have been reported widely in literature, but not always with very clear descriptions and illustrations. We have tried to be as comprehensive as possible in the list below, but we have omitted taxa that we are unable to evaluate from the published information or that we would only be able to confirm by re-examining specimens. As the precise classification of several taxa is still the subject of debate, we have listed the species alphabetically within each phlyum.

Algae

Fuxianospira gyrata Chen & Zhou, 1997
Megaspirellus houi Chen & Erdtmann, 1991
Sinocylindra yunnanensis Chen & Erdtmann, 1991
Yuknessia sp. of Chen & Erdtmann, 1991

Phylum Porifera

Allantospongia mica Rigby & Hou, 1995
Choia xiaolantianensis Hou *et al.*, 1999
Choiaella radiata Rigby & Hou, 1995
Leptomitella confusa Chen, Hou & Lu, 1989
Leptomitella conica Chen, Hou & Lu, 1989
Leptomitus teretiusculus Chen, Hou & Lu, 1989
Paraleptomitella dictyodroma Chen, Hou & Lu, 1989
Paraleptomitella globula Chen, Hou & Lu, 1989
Quadrolaminiella crassa Chen, Hou & Li, 1990
Quadrolaminiella diagonalis Chen, Hou & Li, 1990
Saetaspongia densa Mehl & Reitner *in* Steiner *et al.*, 1993
Sinoflabrum antiquum Zhang & Babcock, 2001 (? Porifera)
Triticispongia diagonata Mehl & Reitner *in* Steiner *et al.*, 1993

Phylum Cnidaria

Priscapennamarina angusta Zhang & Babcock, 2001 (? Cnidaria)
Xianguangia sinica Chen & Erdtmann, 1991

Phylum Ctenophora

Maotianoascus octonarius Chen & Zhou, 1997
Sinoascus papillatus Chen & Zhou, 1997

Phylum Nematomorpha

Cricocosmia jinningensis Hou & Sun, 1988
Maotianshania cylindrica Sun & Hou, 1987
Palaeoscolex sinensis Hou & Sun, 1988

Phylum Priapulida

Acosmia maotiania Chen & Zhou, 1997
Archotuba conoidalis Hou *et al.*, 1999 (? = *Selkirkia*? *elongata* Luo & Hu *in* Luo *et al.*, 1999)
Corynetis brevis Luo & Hu *in* Luo *et al.*, 1999
Palaeopriapulites parvus Hou *et al.*, 1999
Paraselkirkia jinningensis Hou *et al.*, 1999 (? = *Selkirkia sinica* Luo & Hu *in* Luo *et al.*, 1999)

Protopriapulites haikouensis Hou *et al.*, 1999 (= *Sicyophorus rarus* Luo & Hu *in* Luo *et al.*, 1999)

Phylum Chaetognatha

Eognathacantha ercainella Chen & Huang, 2002

Phylum Hyolitha

Ambrolinevitus maximus Jiang, 1982
Ambrolinevitus ventricosus Qian, 1978
Burithes yunnanensis Hou *et al.*, 1999 (? = *Glossolites magnus* Luo & Hu *in* Luo *et al.*, 1999)
Linevitus opimus Yu, 1974

Phylum Annelida?

Polychaete? sp. nov. of Chen & Zhou, 1997

Phylum Lobopodia

Cardiodictyon catenulum Hou, Ramsköld & Bergström, 1991
Hallucigenia fortis Hou & Bergström, 1995
Luolishania longicruris Hou & Chen, 1989
Microdictyon sinicum Chen, Hou & Lu, 1989
Onychodictyon ferox Hou, Ramsköld & Bergström, 1991
Paucipodia inermis Chen, Zhou & Ramsköld, 1995

Anomalocarididae (phylum uncertain)

Amplectobelua symbrachiata Hou, Bergström & Ahlberg, 1995
Anomalocaris saron Hou, Bergström & Ahlberg, 1995
Anomalocaris sp. of Hou, Bergström & Ahlberg, 1995
Cucumericrus decoratus Hou, Bergström & Ahlberg, 1995
Parapeytoia yunnanensis Hou, Bergström & Ahlberg, 1995

Phylum Arthropoda

Acanthomeridion serratum Hou, Chen & Lu, 1989
Almenia spinosa Hou & Bergström, 1997
Branchiocaris? yunnanensis Hou, 1987
Canadaspis laevigata (Hou & Bergström, 1991) (= *Canadaspis eucallus* Chen & Zhou, 1997; = *Perspicaris?* sp. of Hou 1987; ? = *Yiliangocaris ellipticus* Luo & Hu *in* Luo *et al.*, 1999)
Chengjiangocaris longiformis Hou & Bergström, 1991 (*Cambrofengia yunnanensis* Hou *et al.*, 1999 may be detached appendages)
Cindarella eucalla Chen, Ramsköld, Edgecombe & Zhou *in* Chen *et al.*, 1996
Clypecaris pteroidea Hou, 1999 (? = *Ercaicunia multinodosa* Luo & Hu *in* Luo *et al.*, 1999)
Combinivalvula chengjiangensis Hou, 1987
Comptaluta inflata (Zhang, 1974)
Comptaluta leshanensis (Lee, 1975)
Dongshanocaris foliiformis (Hou & Bergström, 1998)
Eoredlichia intermedia (Lu, 1940)
Ercaia minuscula Chen, Vannier & Huang, 2001
Forfexicaris valida Hou, 1999
Fortiforceps foliosa Hou &Bergström, 1997
Fuxianhuia protensa Hou, 1987
Isoxys auritus (Jiang, 1982)
Isoxys curvirostratus Vannier & Chen, 2000
Isoxys paradoxus Hou, 1987 (? = *Isoxys elongatus* Luo & Hu *in* Luo *et al.*, 1999)
Jianfengia multisegmentalis Hou, 1987
Jiucunella paulula Hou & Bergström, 1991
Kuamaia lata Hou, 1987
Kuamaia muricata Hou & Bergström, 1997
Kuanyangia pustulosa (Lu, 1941)
Kunmingella douvillei (Mansuy, 1912)
Kunyangella cheni Huo, 1965
Leanchoilia illecebrosa (Hou, 1987) (= *Dianchia mirabilis* Luo & Hu *in* Luo *et al.*, 1997; = *Leanchoilia asiatica* Luo & Hu *in* Luo *et al.*, 1997; = *Yohoia sinensis* Luo & Hu *in* Luo *et al.*, 1997; ? = *Zhongxinia speciosa* Luo & Hu *in* Luo *et al.*, 1997; ? = *Apiocephalus elegans* Luo & Hu *in* Luo *et al.*, 1999)
Liangshanella liangshanensis Huo, 1956
Naraoia longicaudata Zhang & Hou, 1985
Naraoia spinosa Zhang & Hou, 1985
Occacaris oviformis Hou, 1999

Odaraia? eurypetala Hou & Sun, 1988 (? = *Glossocaris oculatus* Luo & Hu *in* Luo *et al.*, 1999)
Parapaleomerus sinensis Hou *et al.*, 1999
Pectocaris spatiosa Hou, 1999
Pisinnocaris subconigera Hou & Bergström, 1998 (? = *Jianshania furcatus* Luo & Hu *in* Luo *et al.*, 1999)
Pseudoiulia cambriensis Hou & Bergström, 1998
Pygmaclypeatus daziensis Zhang, Han & Shu, 2000
Retifacies abnormalis Hou, Chen & Lu, 1989 (= *Retifacies longispinus* Luo & Hu *in* Luo *et al.*, 1997; = *Tuzoia* sp. of Shu, 1990)
Rhombicalvaria acantha Hou, 1987
Saperion glumaceum Hou, Ramsköld & Bergström, 1991
Sidneyia sinica Zhang & Shu *in* Zhang, Han & Shu, 2002
Sinoburius lunaris Hou, Ramsköld & Bergström, 1991
Skioldia aldna Hou & Bergström, 1997
Squamacula clypeata Hou & Bergström, 1997
Tanglangia longicaudata Luo & Hu *in* Luo *et al.*, 1999
Tsunyidiscus aclis Zhou *in* Lee *et al.*, 1975
Tsunyiella diandongensis Tong *in* Huo & Shu, 1985
?*Tuzoia sinensis* Pan, 1957 (? = *Tuzoia limba* Shu, 1990)
Urokodia aequalis Hou, Chen & Lu, 1989
Waptia ovata (Lee, 1975)
Wutingaspis tingi Kobayashi, 1944
Wutingella binodosa Zhang, 1974
Xandarella spectaculum Hou, Ramsköld & Bergström, 1991
Yunnanocaris megista Hou, 1999
Yunnanocephalus yunnanensis (Mansuy, 1912)

Phylum Phoronida

Iotuba chengjiangensis Chen & Zhou, 1997

Phylum Brachiopoda

Diandongia pista Rong, 1974
Heliomedusa orienta Sun & Hou, 1987
Lingulella chengjiangensis Jin, Hou & Wang, 1993
Lingulellotreta malongensis (Rong, 1974)
Longtancunella chengjiangensis Hou *et al.*, 1999

Phylum? Vetulicolia

Banffia confusa Chen & Zhou, 1997 (? = *Heteromorphus longicaudatus* Luo & Hu *in* Luo *et al.*, 1999)
Didazoon haoae Shu & Han *in* Shu *et al.*, 2001
Pomatrum ventralis Luo & Hu *in* Luo *et al.*, 1999
Vetulicola cuneata Hou, 1987 (? = *Vetulicola rectangulata* Luo & Hu *in* Luo *et al.*, 1999)
Xidazoon stephanus Shu, Conway Morris & Zhang *in* Shu *et al.*, 1999

Phylum Chordata

Cathaymyrus diadexus Shu, Conway Morris & Zhang, 1996
?*Cathaymyrus haikouensis* Luo & Hu *in* Luo *et al.*, 2001
Myllokunmingia fengjiaoa Shu, Zhang & Han *in* Shu *et al.*, 1999 (=*Haikouichthys ercaicunensis* Luo, Hu & Shu *in* Shu *et al.*, 1999)
?*Zhongxiniscus intermedius* Luo & Hu *in* Luo *et al.*, 2001

Enigmatic animals

Allonnia phrixothrix Bengtson & Hou, 2001 (= *Allonnia junyuani* Janussen *et al.*, 2002)
Batofasciculus ramificans Hou *et al.*, 1999
Cotyledion tylodes Luo & Hu *in* Luo *et al.*, 1999 (? = *Cambrotentacus sanwuia* Zhang & Shu *in* Zhang *et al.*, 2001)
Dinomischus venustus Chen, Hou & Lu, 1989
Eldonia eumorpha (Sun & Hou, 1987) (= *Yunnanomedusa eleganta* Sun & Hou 1987)
Facivermis yunnanicus Hou & Chen, 1989
Jiucunia petalina Hou *et al.*, 1999
Maanshania crusticeps Hou *et al.*, 1999
Parvulonoda dubia Rigby & Hou, 1995

Phlogites longus Luo & Hu *in* Luo *et al.*, 1999 (=*Phlogites brevis* Luo & Hu *in* Luo *et al.*, 1999; ? = *Calathites spinalis* Luo & Hu *in* Luo *et al.*, 1999; ? = *Cheunkongella ancestralis* Shu *et al.*, 2001)
Rotadiscus grandis Sun & Hou, 1987
Yunnanozoon lividum Hou, Ramsköld & Bergström, 1991 (? = *Haikouella lanceolata* Chen *et al.*, 1999; ? = *Haikouella jianshanensis* Shu *et al.*, 2003)

Note: While this book was in press, an additional report (Chen *et al.* 2002) named and illustrated nine new species including arthropods, a lobopodian, "worms", a purported chaetognath, and an enigmatic form. The report bears the imprint 2002, but was not generally available until mid 2003.

REFERENCES

Aguinaldo, A.M.A., Turbeville, J.M., Linford, L.S., Rivera, M.C., Garey, J.R., Raff, R.A. & Lake, J.A. 1997. Evidence for a clade of nematodes, arthropods and other animals. *Nature*, **387**, 489–492.

Babcock, L.E. & Zhang Wen-tang. 1997. Comparative taphonomy of two nonmineralized arthropods: *Naraoia* (Nektaspida; Early Cambrian, Chengjiang biota, China) and *Limulus* (Xiphosurida; Holocene, Atlantic Ocean). *Bulletin of the National Museum of Natural Science*, **10**, 233–250.

Babcock, L.E. & Zhang Wen-tang. 2001. Stratigraphy, paleontology and depositional setting of the Chengjiang Lagerstätte (lower Cambrian), Yunnan, China. *In*: Peng Shan-chi, Babcock, L.E. & Zhu Mao-yan (eds), *Cambrian System of South China*, 66–86. University of Science and Technology of China Press, Hefei.

Babcock, L.E., Zhang Wen-tang & Leslie, S.A. 2001. The Chengjiang biota: record of the early Cambrian diversification of life and clues to exceptional preservation of fossils. *GSA Today*, **11**, 4–9.

Bauld, J., D'Amelio, E. & Farmer, J.D. 1992. Modern microbial mats. *In*: Schopf, J.W. & Klein, C. (eds), *The Proterozoic Biosphere*, 262–269. Cambridge University Press, Cambridge.

Beecher, C.E. 1901. Discovery of eurypterid remains in the Cambrian of Missouri. *American Journal of Science*, **12**, 364–366.

Bengtson, S. & Hou Xian-guang. 2001. The integument of Cambrian chancelloriids. *Acta Palaeontologica Polonica*, **46**, 1–22.

Bengtson, S., Matthews, S.C. & Missarzhevsky, V.V. 1981. *In*: Missarzhevsky, V.V. & Mambetov, A.M., Stratigrafiya i fauna pogranichnykh sloev kembriya i dokembriya Malogo Karatau [Stratigraphy and fauna of the Precambrian-Cambrian boundary beds of Malyj Karatau]. *Trudy Geologicheskogo Instituta AN SSSR*, **326**, 90 pp. [In Russian].

Bergström, J. 1986. *Opabinia* and *Anomalocaris*, unique Cambrian 'arthropods'. *Lethaia*, **19**, 241–246.

Bergström, J. 1987. The Cambrian *Opabinia* and *Anomalocaris*. *Lethaia*, **20**, 187–188.

Bergström, J. 1991. Metazoan evolution around the Precambrian-Cambrian transition. *In*: Simonetta, A.M. & Conway Morris, S. (eds), *The Early Evolution of Metazoa and the Significance of Problematic Taxa*, 25–34. Cambridge University Press, Cambridge.

Bergström, J. 1994. Ideas on early animal evolution. *In*: Bengtson, S. (ed.), Early life on Earth. *Nobel Symposium*, **84**, 460–466. Columbia University Press, New York.

Bergström, J. 1997. Origin of high-rank groups of organisms. *Paleontological Research*, **1**, 1–14.

Bergström, J. 2001. Chengjiang. *In*: Briggs, D.E.G. & Crowther, P.R. (eds), *Palaeobiology II: a synthesis*, 241–246. Blackwell Scientific, Oxford.

Bergström, J. & Hou Xian-guang. 1998. Chengjiang arthropods and their bearing on early arthropod evolution. *In*: Edgecombe, G.D. (ed.), *Arthropod Fossils and Phylogeny*, 151–184. Columbia University Press, New York.

Bergström, J. & Hou Xian-guang. 2001. Cambrian Onychophora or Xenusians. *Zoologischer Anzieger*, **240**, 237–245.

Bergström, J., Naumann, W.W., Viehweg, J. & Martí Mus, M. 1998. Conodonts, calcichordates and the origin of vertebrates. *Mitteilungen der Museum für Naturkunde zu Berlin, Geowissenschaftliche Reihe*, **1**, 81–92.

Boxshall, G. 1998. Comparative limb morphology in major crustacean groups: the coxa-basis joint in post-mandibular limbs. *In*: Fortey, R.A. & Thomas, R. (eds), Arthropod relationships. *Systematics Association Special Volume*, **55**, 155–167. Chapman & Hall, London.

Brasier, M.D., Green, O.R., Jephcoat, A.P., Kleppe, A.K., Van Kranendonk, M.J., Lindsay, J.F., Steele, A. & Grassineau, N.V. 2002. Questioning the evidence for Earth's oldest fossils. *Nature*, **416**, 76–81.

Briggs, D.E.G. 1976. The arthropod *Branchiocaris* n. gen., Middle Cambrian, Burgess Shale, British Columbia. *Geological Survey of Canada Bulletin*, **264**, 1–29.

Briggs, D.E.G. 1992. Phylogenetic significance of the Burgess Shale crustacean *Canadaspis*. *Acta Zoologica (Stockholm)*, **73**, 293–300.

Briggs, D.E.G. 1994. Giant predators from the Cambrian of China. *Science*, **264**, 1283–1284.

Briggs, D.E.G. & Collins, D. 1988. A Middle Cambrian chelicerate from Mount Stephen, British Columbia. *Palaeontology*, **31**, 779–798.

Briggs, D.E.G. & Collins, D. 1999. The arthropod *Alalcomenaeus cambricus* Simonetta, from the Middle Cambrian Burgess Shale of British Columbia. *Palaeontology*, **42**, 953–977.

Briggs, D.E.G., Erwin, D.H. & Collier, F.J. 1994. *Fossils of the Burgess Shale*. 238 pp. Smithsonian Institution Press, Washington & London.

Brocks, J.J., Logan, G.A., Buick, R. & Summons, R.E. 1999. Archean molecular fossils and the early rise of eukaryotes. *Science*, **285**, 1033–1036.

Budd, G.E. 1996. The morphology of *Opabinia regalis* and the reconstruction of the arthropod stem-group. *Lethaia*, **29**, 1–14.

Budd, G.E. 1998. Stem group arthropods from the Lower Cambrian Sirius Passet fauna of North Greenland. *In*: Fortey, R.A. & Thomas, R. (eds), Arthropod relationships. *Systematics Association Special Volume*, **55**, 125–138. Chapman & Hall, London.

Budd, G.E. 2002. A palaeontological solution to the arthropod head problem. *Nature*, **417**, 271–275.

Budd, G.E., Butterfield, N.J. & Jensen, S. 2001. Crustaceans and the "Cambrian Explosion". *Science*, **249**, 2047.

Budd, G.E. & Jensen, S. 2000. A critical reappraisal of the fossil record of the bilaterian phyla. *Biological Reviews*, **75**, 253–295.

Butterfield, N.J. 2002. *Leanchoilia* guts and the interpretation of three-dimensional structures in Burgess Shale-type fossils. *Paleobiology*, **28**, 155–171.

Caster, K.E. & Macke, W.B. 1952. An aglaspid merostome from the Upper Ordovician of Ohio. *Journal of Paleontology*, **26**, 753–757.

Chen Jun-yuan, Dzik, J., Edgecombe, G.D., Ramsköld, L. & Zhou Gui-qing. 1995a. A possible Early Cambrian chordate. *Nature*, **377**, 720–722.

Chen Jun-yuan, Edgecombe, G.D. & Ramsköld, L. 1997. Morphological and ecological disparity in naraoiids (Arthropoda) from the Early Cambrian Chengjiang fauna, China. *Records of the Australian Museum*, **49**, 1–24.

Chen Jun-yuan, Edgecombe, G.D., Ramsköld, L. & Zhou Gui-qing. 1995b. Head segmentation in Early Cambrian *Fuxianhuia*: implications for arthropod evolution. *Science*, **268**, 1339–1343.

Chen Jun-yuan & Erdtmann, B.D. 1991. Lower Cambrian lagerstätte from Chengjiang, Yunnan, China: Insights for reconstructing early metazoan life. *In*: Simonetta, A.M. & Conway Morris, S. (eds), *The Early Evolution of Metazoa and the Significance of Problematic Taxa*, 57–76. Cambridge University Press, Cambridge.

Chen Jun-yuan, Hou Xian-guang & Li Guo-xiang. 1990. New Lower Cambrian demosponges—*Quadrolaminiella* gen. nov. from Chengjiang, Yunnan. *Acta Palaeontologica Sinica*, **29**, 402–414. [In Chinese, with English summary].

Chen Jun-yuan, Hou Xian-guang & Lu Hao-zhi. 1989a. Early Cambrian netted scale-bearing worm-like sea animal. *Acta Palaeontologica Sinica*, **28**, 1–16. [In Chinese, with English summary].

Chen Jun-yuan, Hou Xian-guang & Lu Hao-zhi. 1989b. Lower Cambrian leptomitids (Demospongea), Chengjiang, Yunnan. *Acta Palaeontologica Sinica*, **28**, 17–31. [In Chinese, with English summary].

Chen Jun-yuan, Hou Xian-guang & Lu Hao-zhi. 1989c. Early Cambrian hock glass-like rare sea animal *Dinomischus* (Entoprocta) and its ecological features. *Acta Palaeontologica Sinica*, **28**, 57–81. [In Chinese, with English summary].

Chen Jun-yuan & Huang Di-ying. 2002. A possible Lower Cambrian chaetognath (arrow worm). *Science*, **298**, 187.

Chen Jun-yuan, Huang Di-ying & Li Chia-wei. 1999. An early Cambrian craniate-like chordate. *Nature*, **402**, 518–522.

Chen Jun-yuan & Li Chia-wei. 1997. Early Cambrian chordate from Chengjiang, China. *Bulletin of the National Museum of Natural Science*, **10**, 257–273.

Chen Jun-yuan, Ramsköld, L. & Zhou Gui-qing. 1994. Evidence for monophyly and arthropod affinity of Cambrian giant predators. *Science*, **264**, 1304–1308.

Chen Jun-yuan & Vannier, J. 2000. Building-up of complex marine foodwebs: new fossil evidence from the Early Cambrian Maotianshan Shale biota. The Palaeontological Association Annual Meeting 2000, Edinburgh, Abstracts.

Chen Jun-yuan, Vannier, J. & Huang Di-ying. 2001. The origin of early crustaceans: new evidence from the early Cambrian of China. *Proceedings of the Royal Society, London* B, **268**, 2181–2187.

Chen Jun-yuan & Zhou Gui-qing. 1997. Biology of the Chengjiang fauna. *Bulletin of the National Museum of Natural Science*, **10**, 11–106.

Chen Jun-yuan, Zhou Gui-qing & Ramsköld, L. 1995c. A new Early Cambrian onychophoran-like animal, *Paucipodia* gen. nov., from the Chengjiang fauna, China. *Transactions of the Royal Society of Edinburgh: Earth Sciences*, **85**, 275–282.

Chen Jun-yuan, Zhou Gui-qing & Ramsköld, L. 1995d. The Cambrian lobopodian *Microdictyon sinicum* and its broader significance. *Bulletin of the National Museum of Natural Science*, **5**, 1–93.

Chen Jun-yuan, Zhou Gui-qing, Zhu Mao-yan & Yeh Kuei-yu. 1996. *The Chengjiang Biota. A unique window of the Cambrian explosion*. 222 pp. National Museum of Natural Science, Taichung, Taiwan. [In Chinese].

Chen Jun-yuan, Zhu Mao-yan & Zhou Gui-qing. 1995e. The earliest Cambrian medusiform *Eldonia* from the Chengjiang Lagerstätte. *Acta Palaeontologica Polonica*, **40**, 213–244.

Chen Liang-zhong, Luo Hui-lin, Hu Shi-xue, Yin Gi-yun, Jiang Zhi-wen, Wu Zhi-liang, Li Feng & Chen Ai-lin. 2002. *Early Cambrian Chengjiang Fauna in Eastern Yunnan, China*, 199 pp, 28 pls. Yunnan Science and Technology Press, Kunming. [In Chinese, with English summary].

Chlupáč, I. 1995. Lower Cambrian arthropods from the Paseky Shale (Barrandian area, Czech Republic). *Journal of the Czech Geological Society*, **40**, 9–36.

Collins, D. 1996. The "evolution" of *Anomalocaris* and its classification in the arthropod Class Dinocarida (nov.) and Order Radiodonta (nov.). *Journal of Paleontology*, **70**, 280–293.

Conway Morris, S. 1977a. Fossil priapulid worms. *Special Papers in Palaeontology*, **20**, 1–95.

Conway Morris, S. 1977b. A new metazoan from the Burgess Shale of British Columbia. *Palaeontology*, **20**, 623–640.

Conway Morris, S. 1977c. A new entoproct-like organism from the Burgess Shale of British Columbia. *Palaeontology*, **20**, 833–845.

Conway Morris, S. 1997a. The cuticular structure of the 495-Myr-old type species of the fossil worm *Palaeoscolex*, *P. piscatorum* (?Priapulida). *Zoological Journal of the Linnean Society*, **119**, 69–82.

Conway Morris, S. 1997b. Defusing the Cambrian 'explosion'? *Current Biology*, **7**, R71–74.

Conway Morris, S. 1998. *The Crucible of Creation. The Burgess Shale and the rise of animals*, 242 pp. Oxford University Press, Oxford.

Conway Morris, S. 2000a. The Cambrian explosion: slow-fuse or megatonnage? *Proceedings of the National Academy of Sciences*, **97**, 4426–4429.

Conway Morris, S. 2000b. Nipping the Cambrian "explosion" in the bud? *BioEssays*, **22**, 1053–1056.

Conway Morris, S. & Collins, D.H. 1996. Middle Cambrian ctenophores from the Stephen Formation, British Columbia. *Philosophical Transactions of the Royal Society of London* B, **351**, 279–308.

Conway Morris, S. & Peel, J.S. 1995. Articulated halkieriids from the Lower Cambrian of North Greenland and their role in early Cambrian protostome evolution. *Philosophical Transactions of the Royal Society of London* B, **347**, 305–358.

Conway Morris, S. & Robison, R.A. 1986. Middle Cambrian priapulids and other soft-bodied fossils from Utah and Spain. *The University of Kansas Paleontological Contributions*, **117**, 1–22.

Conway Morris, S. & Robison, R.A. 1988. More soft-bodied animals and algae from the Middle Cambrian of Utah and British Columbia. *The University of Kansas Paleontological Contributions*, **122**, 1–48.

Dahl, E. 1984. The subclass Phyllocarida (Crustacea) and the status of some early fossils: a neontologist's view. *Videnskabelige Meddelelser fre Dansk naturhistorik Forening*, **145**, 61–76.

Delle Cave, L., Insom, E. & Simonetta, A.M. 1998. Advances, divisions, possible relapses and additional problems in understanding the early evolution of the Articulata. *Italian Journal of Zoology*, **65**, 19–38.

Delle Cave, L. & Simonetta, A.M. 1991. Early Paleozoic arthropods and problems of arthropod phylogeny; with notes on taxa of doubtful affinities. *In*: Simonetta, A.M. & Conway Morris, S. (eds), *The Early Evolution of Metazoa and the Significance of Problematic Taxa*, 189–244. Cambridge University Press, Cambridge.

Deprat, J. 1912. Pt. 1, Geologie générale [Text and Atlas]. *In*: Deprat, J. & Mansuy, H., Etude géologique du Yun-Nan oriental. *Mémoires du service géologique de l'Indochine*, **1**, viii. 370 pp, 20 pls + Atlas (178 figs, 7 folded maps).

Dewell, R.A. 2000. Colonial origin for Eumetazoa: major morphological transitions and the origin of bilaterian complexity. *Journal of Morphology*, **243**, 35–74.

Dunlop, J.A. & Selden, P.A. 1998. The early history and phylogeny of the chelicerates. *In*: Fortey, R.A. & Thomas, R. (eds), Arthropod relationships. *Systematics Association Special Volume*, **55**, 221–235. Chapman & Hall, London.

Dzik, J. 1991. Is fossil evidence consistent with traditional views of the early metazoan phylogeny? *In*: Simonetta, A.M. & Conway Morris, S. (eds), *The Early*

Evolution of Metazoa and the Significance of Problematic Taxa, 47–56. Cambridge University Press, Cambridge.

Dzik, J. 1995. *Yunnanozoon* and the ancestry of chordates. *Acta Palaeontologica Polonica*, **40**, 341–360.

Dzik, J. 2003. Early Cambrian lobopodian sclerites and associated fossils from Kazakhstan. *Palaeontology*, **46**, 93–112.

Dzik, J. & Lendzion, K. 1988. The oldest arthropods of the East European Platform. *Lethaia*, **21**, 29–38.

Dzik, J., Zhao Yuan-long & Zhu Mao-yan. 1997. Mode of life of the Middle Cambrian eldonioid lophophorate *Rotadiscus*. *Palaeontology*, **40**, 385–396.

Edgecombe, G.D. (ed.). 1998. *Arthropod Fossils and Phylogeny*, 347 pp. Columbia University Press, New York.

Edgecombe, G.D. & Ramsköld, L. 1999a. Response to Wills's paper, "Classification of the arthropod *Fuxianhuia*". *Science*, **272**, 747–748.

Edgecombe, G.D. & Ramsköld, L. 1999b. Relationships of Cambrian Arachnata and the systematic position of Trilobita. *Journal of Paleontology*, **73**, 263–287.

Fedo, C.M. & Whitehouse, M.J. 2002. Metasomatic origin of quartz-pyroxene rock, Akilia, Greenland, and implications for Earth's earliest life. *Science*, **296**, 1448–1452.

Feng Wei-min, Mu Xi-nan & Kouchinsky, A.V. 2001. Hyolith-type microstructure in a mollusc-like fossil from the Early Cambrian of Yunnan, China. *Lethaia*, **34**, 303–308.

Fortey, R.A. 1990. Ontogeny, hypostome attachment and trilobite classification. *Palaeontology*, **33**, 529–576.

Fortey, R.A. 2001. The Cambrian explosion exploded? *Science*, **293**, 438–439.

Fortey, R.A., Briggs, D.E.G. & Wills, M.A. 1996. The Cambrian evolutionary 'explosion': decoupling cladogenesis from morphological disparity. *Biological Journal of the Linnaean Society*, **57**, 13–33.

Fortey, R.A., Briggs, D.E.G. & Wills, M.A. 1997. The Cambrian evolutionary 'explosion' recalibrated. *BioEssays*, **19**, 429–434.

Fortey, R.A. & Owens, R.M. 1999. Feeding habits in trilobites. *Palaeontology*, **42**, 429–465.

Fortey, R.A. & Theron, J.N. 1994. A new Ordovician arthropod, *Soomaspis* and the agnostid problem. *Palaeontology*, **37**, 841–861.

Fortey, R.A. & Thomas, R. (eds). 1998. Arthropod relationships. *Systematics Association Special Volume*, **55**, 383 pp. Chapman & Hall, London.

Giribet, G., Edgecombe, G.D. & Wheeler, W.D. 2001. Arthropod phylogeny based on eight molecular loci and morphology. *Nature*, **413**, 157–161.

Glaessner, M.F. 1959. Precambrian Coelenterata from Australia, Africa and Engand. *Nature*, **183**, 1472–1473.

Gorjansky, V.J. & Koneva, S.P. 1983. Nizhnekembriyskiye bezzamkovyye brakhiopody khrebta Malyi Karatau (Juzhnyj Kazakhstan). [Lower Cambrian inarticulate brachiopods of the Malyi Karatau Range (southern Kazakhstan)]. *Trudy Instituta Geologii I Geofiziki Sibirskogo otdeleniya AN SSSR*, **541**, 128–138. [In Russian].

Gould, S.J. 1989. *Wonderful Life: the Burgess Shale and the nature of history*, 347 pp. Norton, New York.

Grotzinger, J.P. & Rothman, D.H. 1996. An abiotic model for stromatolite morphogenesis. *Nature*, **383**, 423–425.

Gürich, G. 1932. *Mimetaster* n. gen. (Crust.) statt *Mimaster* Gürich. *Senckenbergiana*, **14**, 193.

Hagadorn, J.W. 2002. Chengjiang: early record of the Cambrian explosion. *In*: Bottjer, D.J., Etter, W., Hagadorn, J.W. & Tang, C.M. (eds), *Exceptional Fossil Preservation: a unique view on the evolution of marine life*, 35–60. Columbia University Press, New York.

Han Tsu-ming & Runnegar, B. 1992. Megascopic eukaryotic algae from the 2.1-billion-year-old Negaunee Iron Formation, Michigan. *Science*, **257**, 232–235.

Hao Yi-chun & Shu De-gan. 1987. The oldest well-preserved Phaeodaria (Radiolaria) from southern Shaanxi. *Geoscience*, **1**, 301–310. [In Chinese, with English summary].

Ho Chun-cun. 1942. Phosphate deposits of Tungshan, Chengjiang, Yunnan. *Bulletin of the Geological Survey of China*, **35**, 97–106. Pepei, Chungking, China. [In Chinese].

Holland, N.D. & Chen Jun-yuan. 2001. Origin and early evolution of the vertebrates: new insights from advances in molecular biology, anatomy, and palaeontology. *BioEssays*, **23**, 142–151.

Holmer, L.E., Popov, L.E., Koneva, S.P. & Rong Jia-yu. 1997. Early Cambrian *Lingulellotreta* (Lingulata, Brachiopoda) from South Kazakhstan (Malyi Karatau Range) and South China (Eastern Yunnan). *Journal of Paleontology*, **71**, 577–584.

Hou Xian-guang. 1987a. Two new arthropods from Lower Cambrian, Chengjiang, eastern Yunnan. *Acta Palaeontologica Sinica*, **26**, 236–256. [In Chinese, with English summary].

Hou Xian-guang. 1987b. Three new large arthropods from Lower Cambrian, Chengjiang, eastern Yunnan. *Acta Palaeontologica Sinica*, **26**, 272–285. [In Chinese, with English summary].

Hou Xian-guang. 1987c. Early Cambrian large bivalved arthropods from Chengjiang, eastern Yunnan. *Acta Palaeontologica Sinica*, **26**, 286–298. [In Chinese, with English summary].

Hou Xian-guang. 1987d. Oldest Cambrian bradoriids from eastern Yunnan. *In*: *Stratigraphy and Palaeontology of Systemic Boundaries in China, Precambrian-Cambrian Boundary* (1), 537–545. Compiled by Nanjing Institute of Geology and Palaeontology, Academia Sinica. Nanjing University Publishing House, Nanjing.

Hou Xian-guang. 1999. New rare bivalved arthropods from the Lower Cambrian Chengjiang fauna, Yunnan, China. *Journal of Paleontology*, **73**, 102–116.

Hou Xian-guang, Aldridge, R.J., Siveter, David J., Siveter, Derek J. & Feng Xiang-hong. 2002a. New evidence on the anatomy and phylogeny of the earliest vertebrates. *Proceedings of the Royal Society, London* B, **269**, 1865–1869.

Hou Xian-guang & Bergström, J. 1991. The arthropods of the Lower Cambrian Chengjiang fauna, with relationships and evolutionary significance. *In*: Simonetta, A.M. & Conway Morris, S. (eds), *The Early Evolution of Metazoa and the Significance of Problematic Taxa*, 179–187. Cambridge University Press, Cambridge.

Hou Xian-guang & Bergström, J. 1994. Palaeoscolecid worms may be nematomorphs rather than annelids. *Lethaia*, **27**, 11–17.

Hou Xian-guang & Bergström, J. 1995. Cambrian lobopodians—ancestors of extant onychophorans? *Zoological Journal of the Linnean Society*, **114**, 3–19.

Hou Xian-guang & Bergström, J. 1997. Arthropods of the Lower Cambrian Chengjiang fauna, southwest China. *Fossils and Strata*, **45**, 116 pp.

Hou Xian-guang & Bergström, J. 1998. Three additional arthropods from the Early Cambrian Chengjiang fauna, Yunnan, southwest China. *Acta Palaeontologica Sinica*, **37**, 395–401.

Hou Xian-guang, Bergström, J. & Ahlberg, P. 1995. *Anomalocaris* and other large animals in the Lower Cambrian Chengjiang fauna of southwest China. *Geologiska Föreningens i Stockholm Förhandlingar*, **117**, 163–183.

Hou Xian-guang, Bergström, J., Wang Hai-feng, Feng Xiang-hong & Chen Ai-lin. 1999. *The Chengjiang Fauna. Exceptionally well-preserved animals from 530 million years ago.* 170 pp. Yunnan Science and Technology Press, Kunming, Yunnan Province, China. [In Chinese, with English summary].

Hou Xian-guang & Chen Jun-yuan. 1989a. Early Cambrian tentacled worm-like animals (*Facivermis* gen. nov.) from Chengjiang, eastern Yunnan. *Acta Palaeontologica Sinica*, **28**, 32–41. [In Chinese, with English summary].

Hou Xian-guang & Chen Jun-yuan. 1989b. Early Cambrian arthropod-annelid intermediate sea animal, *Luolishania* gen. nov. from Chengjiang, Yunnan. *Acta Palaeontologica Sinica*, **28**, 207–213. [In Chinese, with English summary].

Hou Xian-guang, Chen Jun-yuan & Lu Hao-zhi. 1989. Early Cambrian new arthropods from Chengjiang, Yunnan. *Acta Palaeontologica Sinica*, **28**, 42–57. [In Chinese, with English summary].

Hou Xian-guang, Ma Xiao-ya, Zhao Jie & Bergström, J. In press. The lobopodian *Paucipodia inermis* from the Lower Cambrian Chengjiang fauna, Yunnan, China. *Lethaia*.

Hou Xian-guang, Ramsköld, L. & Bergström, J. 1991. Composition and preservation of the Chengjiang fauna—a Lower Cambrian soft-bodied biota. *Zoologica Scripta*, **20**, 395–411.

Hou Xian-guang, Siveter, D.J., Williams, M. & Feng Xiang-hong. 2002b. A monograph of bradoriid arthropods from the Lower Cambrian of southwest China. *Transactions of the Royal Society of Edinburgh: Earth Sciences*, **92** (for 2001), 347–409.

Hou Xian-guang, Siveter, D.J., Williams, M., Walossek, D. & Bergström, J. 1996. Appendages of the arthropod *Kunmingella* from the early Cambrian of China: its bearing on the systematic position of the Bradoriida and the fossil record of the Ostracoda. *Philosophical Transactions of the Royal Society of London* B, **351**, 1131–1145.

Hou Xian-guang & Sun Wen-guo. 1988. Discovery of Chengjiang fauna at Meishucun, Jinning, Yunnan. *Acta Palaeontologica Sinica*, **27**, 1–12. [In Chinese, with English summary].

Huo Shi-cheng. 1956. Brief notes on Lower Cambrian Archaeostraca from Shensi and Yunnan. *Acta Palaeontologia Sinica*, **4**, 425–445. [In Chinese, with English summary].

Huo Shi-cheng. 1965. Additional notes on Lower Cambrian Archaeostraca from Shensi and Yunnan. *Acta Palaeontologia Sinica*, **13**, 291–307. [In Chinese, with English summary].

Huo Shi-cheng & Shu De-gan. 1985. *Cambrian Bradoriida of South China.* 252 pp, 37 pls. Northwest University Press, Xi'an. [In Chinese, with English summary].

Hupé, P. 1953. Contribution à l'étude du Cambrien inférieur et du Précambrien III de l'Anti-Atlas Marocain. *Notes et Mémoires, Service Géologique, Maroc*, **103** (for 1952), 402 pp.

Janussen, D., Steiner, M. & Zhu Mao-yan. 2002. New well-preserved scleritomes of Chancelloriidae from the Early Cambrian Yuanshan Formation (Chengjiang, China) and the Middle Cambrian Wheeler Shale (Utah, USA) and paleobiological implications. *Journal of Paleontology*, **76**, 596–606.

Javaux, E.J., Knoll, A.H. & Walter, M.R. 2001. Morphological and ecological complexity in early eukaryotic ecosystems. *Nature*, **412**, 66–69.

Jiang Zhi-wen. 1982. Small shelly fossils. *In*: Luo Hui-lin, Jiang Zhi-wen, Wu Xi-che, Song Xue-liang & Ouyang Lin (eds), *The Sinian-Cambrian Boundary in Eastern Yunnan, China*, 163–199. People's Publishing House of Yunnan, China. [In Chinese].

Jin Yu-gan, Hou Xian-guang & Wang Hua-yu. 1993. Lower Cambrian pediculate lingulids from Yunnan, China. *Journal of Paleontology*, **67**, 788–798.

Jin Yu-gan & Wang Hua-yu. 1992. Revision of the Lower Cambrian brachiopod *Heliomedusa* Sun & Hou, 1987. *Lethaia*, **25**, 35–49.

Knoll, A.H. 1996. Archaean and Proterozoic paleontology. *In*: Jansonius, J. & McGregor, D.C. (eds), *Palynology: principles and applications*, **1**, 51–80. American Association of Stratigraphical Palynologists Foundation, Tulsa.

Kobayashi, T. 1936. On the *Parabolinella* fauna from Province Jujuy, Argentina with a note on the Olenidae. *Japanese Journal of Geology and Geography*, **13**, 85–102.

Kobayashi, T. 1944. The Cambrian formations in the middle Yangtze Valley and some trilobites contained therein. Miscellaneous notes on the Cambrian-Ordovician geology and paleontology. *Japan Journal of Geology and Geography*, **19**, 1–4.

Lacalli, T. 2002. Vetulicolians—are they deuterostomes? Chordates? *BioEssays*, **24**, 208–211.

Lantenois, H. 1907. Résultats de la mission géologique et minière du Yun-nan méridinal. I: Note sur la géologie et les mines de la région comprise entre Lao-Kay et Yun-nan-Sen. *Annales des Mines*, 1–134.

Lee Yu-wen. 1975. On the Cambrian ostracodes with new material from Sichuan, Yunnan and Shaanxi, China. *Professional Papers on Stratigraphy and Palaeontology*, **2**, 37–72. Geological Publishing House, Beijing. [In Chinese].

Leslie, S.A., Babcock, L.E. & Zhang Wen-tang. 1996. Community composition and taphonomic overprint of the Chengjiang biota (Early Cambrian, China). *In*: Repetski, J. (ed.), *Sixth North American Paleontological Convention Abstracts of Papers*, 237. The Paleontological Society Special Publication, No. 8.

Li Chia-wei, Chen Jun-yuan & Hua Tzun. 1998. Precambrian sponges with cellular structures. *Science*, **279**, 879–882.

Li Guo-xiang & Zhu Mao-yan. 2001. Discrete sclerites of *Microdictyon* (Lower Cambrian) from the Fucheng Section, Nanzheng, South Shaanxi. *Acta Palaeontologica Sinica*, **40** (supplement), 227–235.

Li Yao-xi, Song Li-sheng, Zhou Zhi-qiang & Yang Jing-yao. 1975. *Early Palaeozoic Stratigraphy of Western Part of Dabashan*, 372 pp, 70 pls. Geological Publishing House, Beijing. [In Chinese].

Lindström, M. 1995. The environment of the early Cambrian Chengjiang fauna. *In*: Chen Jun-yuan, Edgecombe, G. & Ramsköld, L. (eds), *International Cambrian Explosion Symposium* (Programme and Abstracts), 17.

Lu Yen-hao. 1940. On the ontogeny and phylogeny of *Redlichia intermedia* Lu (sp. nov.). *Bulletin of the Geological Society of China*, **20**, 333–342. [In Chinese].

Lu Yen-hao. 1941. Lower Cambrian stratigraphy and trilobite fauna of Kunming, Yunnan. *Bulletin of the Geological Society of China*, **21**, 71–90. [In Chinese].

Luo Hui-lin, Hu Shi-xue & Chen Liang-zhong. 2001. New Early Cambrian chordates from Haikou, Kunming. *Acta Geologica Sinica*, **75**, 345–347.

Luo Hui-lin, Hu Shi-xue, Chen Liang-zhong, Zhang Shi-shan & Tao Yong-he. 1999. *Early Cambrian Chengjiang Fauna from Kunming Region, China*, 129 pp, 32 pls. Yunnan Science and Technology Press, Kunming. [In Chinese, with English summary].

Luo Hui-lin, Hu Shi-xue, Zhang Shi-shan & Tao Yong-he. 1997. New occurrence of the early Cambrian Chengjiang fauna from Haikou, Kunming, Yunnan Province. *Acta Geologica Sinica*, **71**, 97–104. [In Chinese, with English summary].

Luo Hui-lin, Jiang Zhi-wen, Wu Xi-che, Song Xue-ling & Ouyang Lin. 1982. *The Sinian-Cambrian Boundary in eastern Yunnan, China*, 265 pp. People's Publishing House of Yunnan, China.

Luo Hui-lin & Zhang Shi-shan. 1986. Early Cambrian vermes and trace fossils from Jinning-Anning region. *Acta Palaeontologica Sinica*, **25**, 303–311. [In Chinese, with English summary].

Maas, A. & Waloszek, D. 2001. Cambrian derivatives of the early arthropod stem lineage, pentostomids, tardigrades and lobophodians—an "Orsten" perspective. *Zoologischer Anzeiger*, **240**, 451–459.

Malinky, J.M. & Berg-Madsen, V. 1999. A revision of Holm's early and early mid-Cambrian hyoliths of Sweden. *Palaeontology*, **42**, 25–65.

Mansuy, H. 1907. Résultats de la mission géologique et minière du Yun-nan méridinal. III: Résultats paléontologiques. *Annales des Mines*, **11**, 447–472.

Mansuy, H. 1912. Pt. 2, Paléontologie. *In*: Deprat, J. & Mansuy, H., Etude géologique du Yun-Nan oriental. *Mémoires du service géologique de l'Indochine*, **1**, 146 pp, 7 pls.

Marek, L. 1966. New hyolithid genera from the Ordovician of Bohemia. *Casopis Narodniho Muzea*, **135**, 89–92.

Martí Mus, M. & Bergström, J. 2002. The skeletomuscular sysytem of hyolithids. Abstract, Society for Integrative and Comparative Biology, Annual Meeting, Anaheim, California, January 2002.

McKerrow, W.S., Scotese, C.R. & Brasier, M.D. 1992. Early Cambrian continental reconstructions. *Journal of the Geological Society of London*, **149**, 599–606.

Mehl, D. 1996. Organization and microstructure of the chancelloriid skeleton: implications for the biomineralization of the Chancelloriidae. *Bulletin de l'Institut océanographique, Monaco, no. spécial*, **14**, 377–385.

Missarzhevsky, V.V. 1969. Descriptions of hyoliths, gastropods, hyolithelminths, camenids, and forms of an obscure systematic position. *In*: Rozanov, A.Y.U., Missarzhevsky, V.V., Volkova, L.G., Krylov, I.N., Keller, B.M., Korolyuk, I.K., Lendzion, K., Mikhnyak, R., Pykhova, N.G. & Sidorov, A.D. Tommotsky jarus i problema nizhenigranitzy Kembriya, 105–175. Academia Nauk SSSR. *Ordena Trudvogo krasnogo znameni Geologiceskij Institut, Trudy*, **206**. [In Russian].

Mojzsis, S.J., Arrhenius, G., McKeegan, K.D., Harrison, T.M., Nutman, A.P. & Friend, C.R.L. 1996. Evidence for life before 3,800 million years ago. *Nature*, **384**, 55–59.

Müller, K.J. 1979. Phosphatocopine ostracodes with preserved appendages from the Upper Cambrian of Sweden. *Lethaia*, **12**, 1–27.

Müller, K.J. & Hinz-Schallreuter, I. 1993. Palaeoscolecid worms from the Middle Cambrian of Australia. *Palaeontology*, **36**, 549–592.

Narbonne, G.M. & Gehling, J.G. 2002. Life after snowball: the oldest complex Ediacaran fossils. *IPC2002, Geological Society of Australia, Abstracts*, **68**, 122–123.

Narbonne, G.M., Kaufman, A.J. & Knoll, A.H. 1994. Integrated chemostratigraphy and biostratigraphy of the Windermere Supergroup, northwestern Canada: implications for Neoproterozoic correlations and the early evolution of animals. *Bulletin of the Geological Society of America*, **106**, 1281–1292.

Nielsen, C. 1995. *Animal Evolution. Interrelationships of the Living Phyla*. 467 pp. Oxford University Press, Oxford, New York & Tokyo.

Nielsen, C. 1998. The phylogenetic position of the Arthropoda. *In*: Fortey, R.A. & Thomas, R. (eds), Arthropod relationships. *Systematics Association Special Volume*, **55**, 11–22. Chapman & Hall, London.

Nielsen, C. 2001. *Animal Evolution. Interrelationships of the Living Phyla*, 2nd edn, 563 pp. Oxford University Press, Oxford.

Novozhilov, N.I. 1960. Podklass Pseudocrustacea. *In*: Orlov, Yu A. (ed.), *Osnovy Paleontologii, Arthropoda, Trilobitomorpha and Crustacea*, 199. Nedra, Moscow.

Pan, Kiang. 1957. On the discovery of Homopoda from South China. *Acta Palaeontologica Sinica*, **5**, 523–526. [In Chinese].

Pillola, G.L. 1990. Lithologie et trilobites du Cambrien inférieur du SW de la Sardaigne (Italie): implications paléobiogéographiques. *Comptes Rendus de l'Académie des Sciences Paris*, **310**, 321–328.

Pompeckj, J.F. 1927. Ein neues Zeugnis uralten Lebens. *Paläontologische Zeitschrift*, **9**, 287–313.

Popov, L.E. & Holmer, L.E. 2000. Craniopsida. *In*: Williams, A. *et al.* (eds), *Treatise on Invertebrate Paleontology, H (Brachiopoda, Revised)*, **2**, 164–168. The Geological Society of America, Inc. and The University of Kansas, Boulder, Colorado and Lawrence, Kansas.

Qian, Yi. 1978. The Early Cambrian hyolithids of central and southwest China and their stratigraphical significance. *Memoir of the Nanjing Institute of Geology and Palaeontology*, **11**, 1–43. [In Chinese].

Qian Yi & Bengtson, S. 1989. Palaeontology and biostratigraphy of the Early Cambrian Meishucunian Stage in Yunnan Province, South China. *Fossils and Strata*, **24**, 156 pp.

Qian Yi, Li Guo-xiang & Zhu Mao-yan. 2001. The Meishucunian Stage and its small shelly fossil sequence in China. *In*: Zhu Mao-Yan, Van Iten, H., Peng Shan-chi & Li Guo-xiang (eds), The Cambrian of South China. *Acta Palaeontologica Sinica*, **40** (supplement), 54–62.

Ramsköld, L. 1992. Homologies in Cambrian Onychophora. *Lethaia*, **25**, 443–460.

Ramsköld, L. & Chen Jun-yuan. 1998. Cambrian lobopodians: morphology and phylogeny. *In*: Edgecombe, G.D. (ed.), *Arthropod Fossils and Phylogeny*, 107–150. Columbia University Press, New York.

Ramsköld, L., Chen Jun-yuan, Edgecombe, G.D. & Zhou Gui-qing. 1996. Preservational folds simulating tergite junctions in tegopeltid and naraoiid arthropods. *Lethaia*, **29**, 15–20.

Ramsköld, L., Chen Jun-yuan, Edgecombe, G.D. & Zhou Gui-qing. 1997. *Cindarella* and the arachnate clade Xandarellida (Arthropoda, Early Cambrian) from China. *Transactions of the Royal Society of Edinburgh: Earth Sciences*, **88**, 19–38.

Ramsköld, L. & Edgecombe, G.D. 1996. Trilobite appendage structure—*Eoredlichia* reconsidered. *Alcheringa*, **20**, 269–276.

Ramsköld, L. & Hou Xian-guang. 1991. New early Cambrian animal and onychophoran affinities of enigmatic metazoans. *Nature*, **351**, 225–228.

Rasmussen, B., Bengtson, S., Fletcher, I.R. & McNaughton, N. 2002. Discoidal impressions and trace-like fossils more than 1,200 million years old. *Science*, **296**, 1112–1115.

Reitner, J. & Mehl, D. 1995. Early Palaeozoic diversification of sponges: new data and evidences. *Geologisch–Palaeontologische Mitteilungen Innsbruck*, **20**, 335–347.

Resser, C.E. 1929. New Lower and Middle Cambrian Crustacea. *Proceedings of the United States National Museum*, **76**, 1–18.

Retallack, G.J. 1994. Were the Ediacaran fossils lichens? *Paleobiology*, **20**, 523–544.

Rigby, J.K. 1986. Sponges of the Burgess Shale (Middle Cambrian), British Columbia. *Palaeontographica Canada*, **2**, 1–105.

Rigby, J.K. & Hou Xian-guang. 1995. Lower Cambrian demosponges and hexactinellid sponges from Yunnan, China. *Journal of Paleontology*, **69**, 1009–1019.

Rong Jia-yu. 1974. Cambrian brachiopods. *In*: *Handbook of Stratigraphy and Palaeontology of Southwest China*, 54. Edited by Nanjing Institute of Geology and Palaeontology, Academia Sinica, 454 pp. Science Press, Beijing. [In Chinese].

Ruppert, E.E. & Barnes, R.D. 1996. *Invertebrate Zoology*, 6th edn, 1056 pp. Harcourt College Publishers, Orlando, USA.

Salter, J.W. 1866. Appendix. On the fossils of North Wales. *Geological Survey of Great Britain Memoir*, **3**, 240–381, pls 1–26.

Schopf, J.W. 1993. Microfossils of the early Archean Apex chert: new evidence of the antiquity of life. *Science*, **260**, 640–646.

Schram, F.R. 1973. Pseudocoelomates and a nemertine from the Illinois Pennsylvanian. *Journal of Paleontology*, **47**, 985–989.

Seilacher, A. 1992. Vendobionta and Psammocorallia: lost constructions of Precambran evolution. *Journal of the Geological Society of London*, **149**, 607–613.

Seilacher, A., Grazhdankin, D. & Legouta, A. 2003. Ediacaran biota: the dawn of animal life in the shadow of great protists. *Paleontological Research*, **7**, 43–54.

Shu De-gan. 1990. *Cambrian and Lower Ordovician Bradoriida from Zhejiang, Hunan and Shaanxi Provinces*, 95 pp. Northwest University Press, Xi'an. [In Chinese, with English summary].

Shu De-gan, Chen Ling, Han Jian & Zhang Xing-liang. 2001a. An Early Cambrian tunicate from China. *Nature*, **411**, 472–473.

Shu De-gan, Conway Morris, S., Han Jian, Chen Ling, Zhang Xing-liang, Zhang Zhi-fei, Liu Hu-qin, Li Yong & Liu Jia-ni. 2001b. Primitive deuterostomes from the Chengjiang Lagerstätte (Lower Cambrian, China). *Nature*, **414**, 419–424.

Shu De-gan, Conway Morris, S., Han Jian, Zhang Zhi-fei, Yasui, K., Janvier, P., Chen Ling, Zhang Xing-liang, Liu Jia-ni, Li Yong & Liu Hu-qin. 2003. Head and backbone of the Early Cambrian vertebrate *Haikouichthys*. *Nature*, **421**, 526–529.

Shu De-gan, Conway Morris, S. & Zhang Xing-liang. 1996. A *Pikaia*-like chordate from the Lower Cambrian of China. *Nature*, **384**, 157–158.

Shu De-gan, Conway Morris, S., Zhang Xiang-liang, Chen Ling, Li Yong & Han Jian. 1999. A pipiscid-like fossil from the Lower Cambrian of south China. *Nature*, **400**, 746–749.

Shu De-gan, Conway Morris, S., Zhang Zhi-fei, Liu Jia-ni, Han Jian, Chen Ling, Zhang Xing-liang, Yasui, K. & Li Yong. 2003. A new species of yunnanozoan with implications for deuterostome evolution. *Science*, **299**, 1380–1384.

Shu De-gan, Geyer, G., Chen Ling & Zhang Xing-liang. 1995a. Redlichiacean trilobites with preserved soft-parts from the Lower Cambrian Chengjiang Fauna (South China). *In*: Geyer, G. & Landing, E. (eds), Morocco 1995, The Lower-Middle Cambrian standard of Western Gondwana. *Beringia Special Issue*, **2**, 203–241.

Shu De-gan, Luo Hui-lin, Conway Morris, S., Zhang Xing-liang, Hu Shi-xue, Chen Ling, Han Jian, Zhu Min, Li Yong & Chen Liang-zhong. 1999. Lower Cambrian vertebrates from south China. *Nature*, **402**, 42–46.

Shu De-gan, Vannier, J., Luo Hui-lin, Chen Liang-zhong, Zhang Xing-liang & Hu Shi-xue. 1999. Anatomy and lifestyle of *Kunmingella* (Arthropoda, Bradoriida) from the Chengjiang fossil Lagerstätte (Lower Cambrian, Southwest China). *Lethaia*, **32**, 279–298.

Shu De-gan, Zhang Xing-liang & Chen Ling. 1996. Reinterpretation of *Yunnanozoon* as the earliest known hemichordate. *Nature*, **380**, 428–430.

Shu De-gan, Zhang Xiang-liang & Geyer, G. 1995b. Anatomy and systematic affinities of the Lower Cambrian bivalved arthropod *Isoxys auritus*. *Alcheringia*, **19**, 333–342.

Siegmund, H. 1997. Microfacies, geochemistry, and genetic aspects of lowermost Cambrian phosphorites of South China. *Bulletin of the National Museum of Natural Science*, **10**, 143–159.

Simonetta, A.M. 1970. Studies on non-trilobite arthropods from the Burgess Shale (Middle Cambrian). *Palaeontographica Italica*, **66**, 35–45.

Simonetta, A.M. & Delle Cave, L. 1975. The Cambrian non-trilobite arthropods from the Burgess Shale of British Columbia. A study of their comparative morphology, taxonomy and evolutionary significance. *Palaeontographica Italica*, **69**, 1–37.

Siveter, D.J., Waloszek, D., Williams, M. & Fortey, R.A. 2001b. Crustaceans and the "Cambrian Explosion". *Science*, **294,** 2047.

Siveter, D.J., Williams, M., Waloszek, D. 2001a. A phosphatocopid crustacean with appendages from the Lower Cambrian. *Science*, **293**, 479–481.

Steiner, M., Mehl, D., Reitner, J. & Erdtmann, B.D. 1993. Oldest entirely preserved sponges and other fossils from the lowermost Cambrian and a new facies reconstruction of the Yangtze Platform (China). *Berliner Geowissenschaften Abhandlungen*, **9**, 293–329.

Størmer, L. 1956. A Lower Cambrian merostome from Sweden. *Arkiv för zoologie*, **9**, 507–514.

Sun Wei-guo & Hou Xian-guang. 1987a. Early Cambrian medusae from Chengjiang, Yunnan, China. *Acta Palaeontologica Sinica*, **26**, 257–271. [In Chinese, with English summary].

Sun Wei-guo & Hou Xian-guang. 1987b. Early Cambrian worms from Chengjiang, Yunnan, China: *Maotianshania* sp. nov. *Acta Palaeontologica Sinica*, **26**, 300–305. [In Chinese, with English summary].

Syssoiev, V. A. 1958. The superorder Hyolithoidea. *In*: Luppov, N.P. & Drushits, V.V. (eds), *Principles of Palaeontology, Mollusca-Cephalopoda*, 184–190. Academii Nauk SSSR. [In Russian].

Tait, N.N. 2001. The Onychophora and Tardigrada. *In*: Anderson, D.T. (ed.), *Invertebrate Zoology*, 2nd edn, 206–224. Oxford University Press, Oxford.

Thompson, I. & Jones, D.S. 1980. A possible onychophoran from the Middle Pennsylvanian Mazon Creek Beds of northern Illinois. *Journal of Paleontology*, **54**, 588–596.

Tong Hao-wen. 1989. A preliminary study on the *Microdictyon* from the Lower Cambrian of Zhenba, South Shaanxi. *Acta Micropalaeontologica Sinica*, **6**, 97–101.

Ulrich, E.O. 1899. Preliminary description of new Lower Silurian sponges. *American Geologist*, **3**, 233–248.

van Zuilen, M.A., Lepland, A. & Arrhenius, G. 2002. Reassessing the evidence for the earliest traces of life. *Nature*, **418**, 627–630.

Vannier, J. & Chen Jun-yuan. 2000. The Early Cambrian colonization of pelagic niches exemplified by *Isoxys* (Arthropoda). *Lethaia*, **33**, 295–311.

Vannier, J. & Chen, Jun-yuan. 2002. Digestive system and feeding mode in Cambrian naraoiid arthropods. *Lethaia*, **35**, 107–120.

Walcott, C.D. 1886. Second contribution to the studies of the Cambrian faunas of North America. *United States Geological Survey Bulletin*, **30**, 369 pp.

Walcott, C.D. 1890. The fauna of the Lower Cambrian or *Olenellus* Zone. *Reports of the U. S. Geological Survey*, **10**, 509–763.

Walcott, C.D. 1911a. Middle Cambrian Merostomata. Cambrian Geology and Paleontology II. *Smithsonian Miscellaneous Collections*, **57**, 17–40.

Walcott, C.D. 1911b. Middle Cambrian Holothurians and Medusae. Cambrian Geology and Paleontology II. *Smithsonian Miscellaneous Collections*, **57**, 41–68.

Walcott, C.D. 1911c. Middle Cambrian Annelids. Cambrian Geology and Paleontology II. *Smithsonian Miscellaneous Collections*, **57**, 109–144.

Walcott, C.D. 1912. Middle Cambrian Branchiopoda, Malacostraca, Trilobita and Merostomata. Cambrian Geology and Paleontology II. *Smithsonian Miscellaneous Collections*, **57**, 145–288.

Walcott, C.D. 1918. Geological explorations in the Canadian Rockies. *In*: Explorations and Fieldwork of the Smithsonian Institution in 1917. *Smithsonian Miscellaneous Collections*, **68**, 4–20.

Walcott, C.D. 1919. Middle Cambrian Algae. Cambrian Geology and Paleontology IV. *Smithsonian Miscellaneous Collections*, **67**, 217–260.

Walcott, C.D. 1920. Middle Cambrian Spongiae. Cambrian Geology and Paleontology IV. *Smithsonian Miscellaneous Collections*, **67**, 261–364.

Walcott, C.D. 1931. Addenda to descriptions of Burgess Shale fossils. *Smithsonian Miscellaneous Collections*, **85**, 1–46 (with explanatory notes by C.E. Resser).

Walossek, D. 1999. On the Cambrian diversity of Crustacea. *In*: Schram, F.R. & Von Vaupel Klein, J.C. (eds), *Crustaceans and the Biodiversity Crisis*, 3–27. Proceedings of the Fourth International Crustacean Congress, Amsterdam, The Netherlands, July 20–24, 1998, **1**. Brill Academic Publishers, Leiden.

Walossek, D. & Müller, K.J. 1998. Early Arthropod phylogeny in light of the Cambrian 'Orsten' fossils. *In*: Edgecombe, G. (ed.), *Arthropod Fossils and Phylogeny*, 185–231. Columbia University Press, New York.

Walter, M.R. 1994. Stromatolites: The main geological source of information on the evolution of the early benthos. *In*: Bengtson, S. (ed.), Early life on Earth. *Nobel Symposium*, **84**, 270–286. Columbia University Press, New York.

Whiteaves, J.F. 1892. Description of a new genus and species of phyllocarid crustacean from the Middle Cambrian of Mount Stephen, British Columbia. *Canadian Records of Science*, **5**, 205–208.

Whittard, W.F. 1953. *Palaeoscolex piscatorum* gen. et sp. nov., a worm from the Tremadocian of Shropshire. *Quarterly Journal of the Geological Society, London*, **109**, 125–135.

Whittington, H.B. & Briggs, D.E.G. 1985. The largest Cambrian animal, *Anomalocaris*, British Columbia. *Philosophical Transactions of the Royal Society of London* B, **309**, 569–609.

Williams, M., Siveter, D.J. & Peel, J. 1996. *Isoxys* (Arthropoda) from the early Cambrian Sirius Passet Lagerstätte, North Greenland. *Journal of Paleontology*, **70**, 947–954.

Wills, M.A. 1996. Classification of the arthropod *Fuxianhuia*. *Science*, **272**, 746–747.

Wills, M.A. 1998a. A phylogeny of recent and fossil Crustacea derived from morphological characters. *In*: Fortey, R.A. & Thomas, R. (eds), Arthropod relationships. *Systematics Association Special Volume*, **55**, 191–209. Chapman & Hall, London.

Wills, M.A. 1998b. Cambrian and Recent disparity: the picture from priapulida. *Paleobiology*, **24**, 177–199.

Wills, M.A., Briggs, D.E.G., Fortey, R.A., Wilkinson, M. & Sneath, P.H.A. 1998. An arthropod phylogeny based on fossil and Recent taxa. *In*: Edgecombe, G.D. (ed.), *Arthropod Fossils and Phylogeny*, 33–105. Columbia University Press, New York.

Wills, M.A. & Fortey, R.A. 2000. The shape of life: how much is written in stone? *BioEssays*, **22**, 1142–1152.

Wray, G.A., Levinton, J.S. & Shapiro, L.H. 1996. Molecular evidence for deep Precambrian divergences among metazoan phyla. *Science*, **274**, 568–573.

Xiao Shu-hai. 2002. Mitotic topologies and mechanics of Neoproterozoic algae and animal embryos. *Paleobiology*, **28**, 244–250.

Xiao Shu-hai, Yuan Xun-lai, Steiner, M. & Knoll, A.H. 2002. Macroscopic carbonaceous compressions in a terminal Proterozoic shale: a systematic reassessment of the Miaohe biota, South China. *Journal of Paleontology*, **76**, 347–376.

Xiao Shu-hai, Zhang Yun & Knoll, A.H. 1998. Three-dimensional preservation of algae and animal embryos in a Neoproterozoic phosphorite. *Nature*, **391**, 553–558.

Yu Wen. 1974. Cambrian hyolithids. *In*: *Handbook of Stratigraphy and Palaeontology of Southwest China*, 111–112. Edited by Nanjing Institute of Geology and Palaeontology, Academia Sinica, 454 pp. Science Press, Beijing. [In Chinese].

Zhang Wen-tang. 1951. Trilobites from the Shipai Shale and their stratigraphical significance. *Huai Shuan* (Newsletter of the Geological Society of China), **2** (for 1950), 10.

Zhang Wen-tang. 1974. Bradoriida. *In*: *Handbook of Stratigraphy and Palaeontology of Southwest China*, 107–111. Edited by Nanjing Institute of Geology and Palaeontology, Academia Sinica, 454 pp. Science Press, Beijing. [In Chinese].

Zhang Wen-tang. 1987a. World's oldest Cambrian trilobites from eastern Yunnan. *In*: *Stratigraphy and Palaeontology of Systemic Boundaries in China, Precambrian-Cambrian Boundary* (1), 537–545. Compiled by Nanjing Institute of Geology and Palaeontology, Academia Sinica. Nanjing University Publishing House, Nanjing.

Zhang Wen-tang. 1987b. Early Cambrian Chengjiang fauna and its trilobites. *Acta Palaeontologica Sinica*, **26**, 223–236. [In Chinese, with English summary].

Zhang Wen-tang & Babcock, L.E. 2001. New extraordinarily preserved enigmatic fossils, possibly with Ediacaran affinities, from the lower Cambrian of Yunnan, China. *Acta Palaeontologica Sinica*, **40** (supplement), 201–213.

Zhang Wen-tang & Hou Xian-guang. 1985. Preliminary notes on the occurrence of the unusual trilobite *Naraoia* in Asia. *Acta Palaeontologica Sinica*, **24**, 591–595. [In Chinese, with English summary].

Zhang Xian-liang, Han Jian & Shu De-gan. 2000. A new arthropod, *Pygmaclypeatus daziensis*, from the Early Cambrian Chengjiang Lagerstätte, South China. *Journal of Paleontology*, **74**, 800–803.

Zhang Xi-guang, Hou Xian-guang & Emig, C.C. In press. Evidence of lophophore diversity in Early Cambrian Brachiopoda. *Proceedings of the Royal Society, London B*.

Zhang Xing-liang, Han Jian & Shu De-gan. 2002. New occurrence of the Burgess Shale arthropod *Sidneyia* in the Early Cambrian Chengjiang Lagerstätte (South China), and revision of the arthropod *Urokodia*. *Alcheringa*, **26**, 1–8.

Zhang Xing-liang, Han Jian, Zhang Zhi-fei, Liu Hu-qin & Shu De-gan. 2003. Reconsideration of the supposed naraoiid larva from the early Cambrian Chengjiang Lagerstätte, South China. *Palaeontology*, **46**, 447–465.

Zhang Xing-liang, Shu De-gan, Li Yong & Han Jian. 2001. New sites of Chengjiang fossils: crucial windows on the Cambrian Explosion. *Journal of the Geological Society of London*, **158**, 211–218.

Zhao Yuan-long, Steiner, M., Yang Rui-dong, Erdtmann, B.D., Guo Qing-jun, Zhou Zhen & Wallis, E. 1999c. Discovery and significance of the early metazoan biotas from the lower Cambrian Niutitang Formation, Zunyi, Guizhou, China. *Acta Palaeontologica Sinica*, **38** (supplement), 132–144. [In Chinese, with English summary].

Zhao Yuan-long, Yuan Jin-liang, Zhu Mao-yan, Yang Rui-dong, Guo Qing-jun, Qian Yi, Huang You-zhuang & Pan Yu. 1999a. A progress report on research on the early Middle Cambrian Kaili biota, Guizhou, P.R.C. *Acta Palaeontologica Sinica*, **38** (supplement), 1–14. [In Chinese, with English summary].

Zhao Yuan-long & Zhu Mao-yan. 1994. Medusiform fossils of Kaili fauna from Taijiang, Guizhou. *Acta Palaeontologica Sinica*, **33**, 272–280. [In Chinese, with English summary].

Zhao Yuan-long, Zhu Mao-yan, Guo Qing-jun & Van Iten Heyo. 1999b. Worms from the Middle Cambrian Kaili biota, Guizhou, P.R.C. *Acta Palaeontologica Sinica*, **38** (supplement), 79–87. [In Chinese, with English summary].

Zhu Mao-yan. 1997. Precambrian-Cambrian trace fossils from eastern Yunnan, China: implications for Cambrian explosion. *Bulletin of the National Museum of Natural Science*, **10**, 275–312.

Zhu Mao-yan, Zhang Jun-ming & Li Guo-xiang. 2001a. Sedimentary environments of the early Cambrian Chengjiang biota: sedimentology of the Yu'anshan Formation in Chengjiang County, eastern Yunnan. *In*: Zhu Mao-yan, Van Iten, H., Peng Shan-chi & Li Guo-xiang (eds), The Cambrian of South China. *Acta Palaeontologica Sinica*, **40** (supplement), 80–105.

Zhu Mao-yan, Zhang Jun-ming & Li Guo-xiang. 2001b. The early Cambrian Chengjiang biota: quarries of non-mineralized fossils at Maotianshan and Ma'anshan, Chengjiang County, Yunnan Province, China. *In*: Peng Shan-chi, Babcock, L.E. & Zhu Mao-yan (eds), *Cambrian System of China*, 219–225. University of Science and Technology of China Press, Hefei.

Zhu Mao-yan, Zhao Yuan-long & Chen Jun-yuan. 2002. Revision of the Cambrian discoidal animals *Stellostomites eumorphus* and *Pararotadiscus guizhouensis* from South China. *Geobios*, **35**, 165–185.

SYSTEMATIC INDEX

Page numbers in *italic* refer to pages with figures

GENERAL INDEX

Page numbers in *italic* refer to pages with figures